Biodiversity and Pest Management in Agroecosystems

Second Edition

FOOD PRODUCTS PRESS®
Crop Science
Amarjit S. Basra, PhD
Senior Editor

Mineral Nutrition of Crops: Fundamental Mechanisms and Implications by Zdenko Rengel

Conservation Tillage in U.S. Agriculture: Environmental, Economic, and Policy Issues by Noel D. Uri

Cotton Fibers: Developmental Biology, Quality Improvement, and Textile Processing edited by Amarjit S. Basra

Heterosis and Hybrid Seed Production in Agronomic Crops edited by Amarjit S. Basra

Intensive Cropping: Efficient Use of Water, Nutrients, and Tillage by S. S. Prihar, P. R. Gajri, D. K. Benbi, and V. K. Arora

Physiological Bases for Maize Improvement edited by María E. Otegui and Gustavo A. Slafer

Plant Growth Regulators in Agriculture and Horticulture: Their Role and Commercial Uses edited by Amarjit S. Basra

Crop Responses and Adaptations to Temperature Stress edited by Amarjit S. Basra

Plant Viruses As Molecular Pathogens by Jawaid A. Khan and Jeanne Dijkstra

In Vitro Plant Breeding by Acram Taji, Prakash P. Kumar, and Prakash Lakshmanan

Crop Improvement: Challenges in the Twenty-First Century edited by Manjit S. Kang

Barley Science: Recent Advances from Molecular Biology to Agronomy of Yield and Quality edited by Gustavo A. Slafer, José Luis Molina-Cano, Roxana Savin, José Luis Araus, and Ignacio Romagosa

Tillage for Sustainable Cropping by P. R. Gajri, V. K. Arora, and S. S. Prihar

Bacterial Disease Resistance in Plants: Molecular Biology and Biotechnological Applications by P. Vidhyasekaran

Handbook of Formulas and Software for Plant Geneticists and Breeders edited by Manjit S. Kang

Postharvest Oxidative Stress in Horticultural Crops edited by D. M. Hodges

Encyclopedic Dictionary of Plant Breeding and Related Subjects by Rolf H. G. Schlegel

Handbook of Processes and Modeling in the Soil-Plant System edited by D. K. Benbi and R. Nieder

The Lowland Maya Area: Three Millennia at the Human-Wildland Interface edited by A. Gómez-Pompa, M. F. Allen, S. Fedick, and J. J. Jiménez-Osornio

Biodiversity and Pest Management in Agroecosystems, Second Edition by Miguel A. Altieri and Clara I. Nicholls

Plant-Derived Antimycotics: Current Trends and Future Prospects edited by Mahendra Rai and Donatella Mares

Concise Encyclopedia of Temperate Tree Fruit edited by Tara Auxt Baugher and Suman Singha

Landscape Agroecology by Paul A. Wojtkowski

Concise Encyclopedia of Plant Pathology by P. Vidhyasekaran

Biodiversity and Pest Management in Agroecosystems

Second Edition

Miguel Angel Altieri, PhD
Clara Ines Nicholls, PhD

CRC Press is an imprint of the
Taylor & Francis Group, an **informa** business

First published 2004 by The Haworth Press, Inc.

Published 2019 by CRC Press
Taylor & Francis Group
6000 Broken Sound Parkway NW, Suite 300
Boca Raton, FL 33487-2742

© 2004 by Taylor & Francis Group, LLC
CRC Press is an imprint of the Taylor & Francis Group, an informa business

No claim to original U.S. Government works

ISBN-13: 978-1-56022-923-0 (pbk)

Cover design by Brooke R. Stiles.

Cover photos of vineyards by Miguel Altieri.

Library of Congress Cataloging-in-Publication Data

Altieri, Miguel A.
    Biodiversity and pest management in agroecosystems / Miguel Angel Altieri, Clara I. Nicholls.— 2nd ed.
        p. cm.
    Includes bibliographical references (p.   ) and index.
    ISBN 1-56022-922-5 (hardcover : alk. paper)—ISBN 1-56022-923-3 (softcover : alk. paper)
    1. Insect pests—Biological control. 2. Insects—Ecology. 3. Biological diversity. 4. Agricultural ecology. I. Nicholls, Clara I. II. Title.
SB933.3 .A38 2003
632'.7—dc21

                                                                                    2002014860

Visit the Taylor & Francis Web site at
http://www.taylorandfrancis.com

and the CRC Press Web site at
http://www.crcpress.com

CONTENTS

ABOUT THE AUTHORS

M. A. Altieri, PhD, has been a Professor of Agroecology at UC Berkeley for more than two decades. He is recognized as a pioneer in developing ecologically based pest management systems, especially through the diversification of agroecosystems. He has published more than 10 books, including *Agroecology: The Science of Sustainable Agriculture,* and more than 200 articles in various journals.

C. I. Nicholls, PhD, is a Research Fellow in the Division of Insect Biology at UC Berkeley. She is an expert in biological control and has conducted considerable research on habitat management strategies to enhance beneficial insects in agricultural systems. She is the author of numerous articles on biological control of insect pests in many international journals.

Preface to the Second Edition

Ten years ago, the writing of this book was a long overdue task. At that time, the pioneering work of researchers such as Helmut van Emden, Robert van den Bosch, David Pimentel, and Richard Root provided the initial enlightenment and inspiration. Much of this work was continued by various colleagues, many of them former University of California at Berkeley graduate students (Deborah K. Letourneau, Matt Liebman, Cliff Gold, David Nestel, Michael Costello) who, with much scientific elegance, have contributed substantially to the advancement of this field.

After considerable debate on the entomological merits of diversification, involving colleagues such as the late Steve Risch, John Vandermeer, David Andow, and Peter Kareiva who cautioned about the universality of the effects of diversity on pest populations and called for more research, there has been a renewed interest in the effects of habitat diversification on insect ecology. This contemporary reevaluation of the field has been championed by a new generation of scientists such as Robert L. Bugg, G. M. Gurr, J. A. Lys, and Doug A. Landis. Despite all of the scientific evidence, years ago we became convinced that diversity in agriculture is not only essential for pest suppression but also crucial to ensure the biological basis for the sustainability of production. We have witnessed the benefits of mimicking nature in our own experimental plots and, more important, in hundreds of farmers' fields in developing countries and in California. Projects led by many nongovernmental organizations (NGOs) in Latin America have used biodiversity as the basis of an agroecological approach tailored to meet the needs of resource-poor farmers through more stable yields and conservation of local resources. These actions have resulted in enhanced food security, reduced use of toxic pesticides, and consequently healthier and better-fed rural families. These testimonies have been significant enough, often more so than what statistical analysis may suggest, to convince us of the benefits of biodiversity in agriculture.

We write the second edition of this book perhaps not so much as entomologists but as agroecologists or, better, as self-made social scientists. This evolution was natural since our research on intercropping placed us face to face with the challenges of rural development in Latin America and the powerful agrarian structure of California. It is a healthy evolution, as environmental problems in agriculture are not only ecological but also part of a social, economic, and political process. Therefore, the root causes of most pest problems affecting agriculture are inherent in the structural features of the prevailing agroeconomic system, which encourages energy-intensive, large-scale, and specialized monocultures. It is our expectation that pest managers will become more sensitive to social, economic, and cultural issues, as pest problems cannot be understood by disentangling ecological from socioeconomic factors. On the contrary, the agroecological paradigm maintains that both must be examined holistically. Only a broader understanding will ensure that the benefits of a diversified agriculture can expand beyond the obvious entomological advantages to include concerns for social equity, economic viability, and cultural compatibility.

Acknowledgments

We are indebted to the hundreds of farmers in Latin America, California, and many other parts of the world who have showed us ecologically sound working farms, which helped us become aware of the importance of diversification in agriculture. This book was initially written while the then-existing Division of Biological Control, University of California, Berkeley, was a world-leading center of research on alternatives to pesticides. During that time, many colleagues and faculty provided encouragement and support, including Linda Schmidt, Javier Trujillo, Jeff Dlott, Andres Yurjevic, Deborah Letourneau, Steve Gliessman, Matt Liebman, Leo Caltagirone, Donald Dahlsten, M. Alice Garcia, Marta Astier, and many others. The Jesse Smith Noyes Foundation of New York provided generous financial support for many years.

Almost ten years after the first edition of this book, it is necessary to thank many additional people for their enthusiasm, faith, and especially dedication for translating many of the ideas in this book into practice: Christos Vasilikiotis, Marco Barzman, Fabian Banga, Peter Rosset, Patrick Archie, Josh Mimer, Raul and Carlos Venegas, Santiago Sarandon, Eddy Peralta, Alfredo Jimenez, the extensionists of EMATER (Instuto de Assistencia e Extensão Rural) and EPAGRI (Empresa de Pesquisa e Extensão Rural de Santa Catarims) in Brazil, the research team of CATIE (Tropical Agricultural Research and Higher Education Center) in Nicaragua, the graduate students of the agroecology master's program at the International University of Andalucia and the Mediterranean Agronomic Institute of Bari, as well as the hundreds of NGO personnel and farmers that constantly strive to enhance biodiversity in the rural landscapes of Latin America.

We are solely responsible for the views expressed in this book and hope that the assembled information provides a useful tool for students and field practitioners to understand and obtain criteria on the diversification of agroecosystems for enhanced pest management.

We dedicate this book to our mothers, families, and friends, as well as all of the current and future farmers of the world, so that they can use the principles of agroecology to better nurture the land and attain a truly sustainable agriculture.

New donors have supported our efforts, so our special gratitude goes to the Foundation for Deep Ecology, the Clarence Heller Charitable Foundation, the Organic Farming Research Foundation, and the California Department of Food and Agriculture (CDFA) Department of Pesticide Regulation. We also thank the Rockefeller Foundation for hosting us at their Bellagio Center where, while contemplating enchanting Lake Como, we wrote much of this second edition.

Introduction

Agriculture implies the simplification of nature's biodiversity and reaches an extreme form in crop monocultures. The end result is the production of an artificial ecosystem requiring constant human intervention. In most cases, this intervention is in the form of agrochemical inputs which, in addition to temporarily boosting yields, result in a number of undesirable environmental and social costs (Altieri, 1987).

As agricultural modernization progresses, ecological principles are continuously ignored or overridden. As a consequence, modern agroecosystems are unstable. Breakdowns manifest themselves as recurrent pest outbreaks in many cropping systems and in the forms of salinization, soil erosion, pollution of water systems, etc. The worsening of most pest problems has been experimentally linked to the expansion of crop monocultures at the expense of vegetation diversity, which is an essential landscape component providing key ecological services to ensure crop protection (Altieri and Letourneau, 1982). Ninety-one percent of the world's 1.5 billion hectares of cropland are under annual crops, mostly monocultures of wheat, rice, maize, cotton, and soybeans. One of the main problems arising from the homogenization of agricultural systems is an increased vulnerability of crops to insect pests and diseases, which can be devastating if they infest a uniform crop, especially in large plantations. To protect these crops worldwide, about 4.7 million pounds of pesticides were applied in 1995 (1.2 billion pounds in the United States); such pesticide injection has increased in the past ten years. In the United States, environmental and social costs associated with such pesticide levels have been estimated at $8 billion per year (Pimentel et al., 1980). Such costs are still valid today. Crop losses due to pests remain at 30 percent, no different from thirty to forty years ago. In California, pesticide use increased from 161 to 212 million pounds of active ingredient, despite the fact that crop acreage remained constant and that research in integrated pest management (IPM) is quite advanced (Liebman, 1997). These are clear signs that the pesticide-based approach to pest control has reached its limits. An alternative approach is needed—one based on the use of ecological principles in order to take full advantage of the benefits of biodiversity in agriculture.

This book analyzes the ecological basis for the maintenance of biodiversity in agriculture and the role it can play in restoring the ecological balance of agroecosystems so that sustainable production may be achieved. Biodiversity performs a variety of renewal processes and ecological services in agroecosystems; when they are lost, the costs can be significant (Altieri, 1991b).

The book focuses particularly on the ways in which biodiversity can contribute to the design of pest-stable agroecosystems. The effects of intercropping, cover cropping, weed management, and crop-field border vegetation manipulation are discussed. A considerable amount of attention is paid to understanding the effects of these vegetationally diverse systems on pest population density and the mechanisms underlying pest reduction in polycultures. This is essential if vegetation management is to be used effectively as the basis of ecologically based pest management (EBPM) tactics in sustainable agriculture.

Although insect communities in agroecosystems can be stabilized by constructing vegetational architectures that support natural enemies and/or directly inhibit pest attack, this book stresses the fact that each situation must be assessed separately, given that long-term vegetation-management strategies are site specific and need to be developed with regard to local and regional environmental, socioeconomic, and cultural factors. In this way, crop mixtures may serve to meet the broader needs and preferences of local farmers and, at the same time, enhance environmental quality.

This book builds on information emerging from renewed interest among scientists in the field of habitat management in enhancing biological control of insects (Barbosa, 1998; Pickett and Bugg, 1998; Landis, Wratten, and Gurr, 2000; Smith and McSorley, 2000).

Chapter 1

The Ecological Role of Biodiversity in Agriculture

Biodiversity refers to all species of plants, animals, and micro-organisms existing and interacting within an ecosystem (McNeely et al., 1990). Global threats to biodiversity should not be foreign to agriculturalists, since agriculture, which covers about 25 to 30 percent of world land area, is perhaps one of the main activities affecting biological diversity. It is estimated that the global extent of cropland increased from around 265 million hectares (ha) in 1700 to around 1.5 billion hectares today, predominantly at the expense of forest habitats (Thrupp, 1997). Very limited areas remain totally unaffected by agriculture-induced land-use changes.

Clearly, agriculture implies the simplification of the structure of the environment over vast areas, replacing nature's diversity with a small number of cultivated plants and domesticated animals (Andow, 1983a). In fact, the world's agricultural landscapes are planted with only some twelve species of grain crops, twenty-three vegetable-crop species, and about thirty-five fruit- and nut-crop species (Fowler and Mooney, 1990); that is no more than seventy plant species which spread over approximately 1,440 million ha of presently cultivated land in the world (Brown and Young, 1990), a sharp contrast with the diversity of plant species found within one hectare of a tropical rain-forest which typically contains over 100 species of trees (Myers, 1984). Of the 7,000 crop species used in agriculture, only 120 are important at a national level. An estimated 90 percent of the world's calorie intake comes from just thirty crops, a small sample of the vast crop diversity available. Monocultures may have temporary economic advantages for farmers, but in the long term they do not represent an ecological optimum (USDA, 1973). Rather, the drastic narrowing of cultivated plant diversity has put the world's food production in greater peril (National Academy of Sciences [NAS], 1972; Robinson, 1996).

The process of environmental simplification reaches an extreme form in agricultural monocultures, which affects biodiversity in various ways:

- Expansion of agricultural land with loss of natural habitats
- Conversion into homogeneous agricultural landscapes with low habitat value for wildlife
- Loss of wild species and beneficial agrodiversity as a direct consequence of agrochemical inputs and other practices
- Erosion of valuable genetic resources through increased use of uniform high-yielding varieties (HYV)

In the developing world, agricultural diversity has been eroded as monocultures dominate. For example, in Bangladesh the promotion of Green Revolution rice led to the loss of diversity, including nearly 7,000 traditional rice varieties and many fish species. Similarly, in the Philippines, the introduction of HYV rice displaced more than 300 traditional rice varieties. In North America similar losses in crop diversity are occurring. Eighty-six percent of the 7,000 apple varieties used in the United States between 1804 and 1904 are no longer in cultivation; of 2,683 pear varieties, 88 percent are no longer available. In Europe thousands of varieties of flax and wheat vanished following the take-over by modern varieties (Thrupp, 1997). In fact, modern agriculture is shockingly dependent on a handful of varieties for its major crops. For example, in the United States, three decades ago 60 to 70 percent of the total bean acreage was planted with two to three bean varieties, 72 percent of the potato acreage with four varieties, and 53 percent of cotton with only three varieties (NAS, 1972).

Researchers have repeatedly warned about the extreme vulnerability associated with this genetic uniformity. Perhaps the most striking example of vulnerability associated with homogenous uniform agriculture was the collapse of Irish potato production in 1845, where the uniform stock of potatoes was highly susceptible to the blight *(Phytophthora infestans)*. During the nineteenth century in France, wine grape production was wiped out by a virulent pest *(Phylloxera vitifoliae)* which eliminated 4 million hectares of uniform grape varieties. Banana monocultural plantations in Costa Rica have been repeatedly seriously jeopardized by diseases such as *Fusarium oxysporum* and yellow Sigatoka. In the United States, in the early 1970s, uniform high-yielding maize hybrids constituted about 70 percent of all the maize varieties; a 15 percent loss of the entire crop by leaf blight occurred in that decade (Adams, Ellingbae, and Rossineau, 1971). Recent expansion of transgenic maize and soybean monocultures, reaching about 45 million hectares mostly grown in the United States in less than six years, represents a worrisome landscape-simplification trend of homogenization (Marvier, 2001).

The net result of biodiversity simplification for agricultural purposes is an artificial ecosystem that requires constant human intervention. Commercial seedbed preparation and mechanized planting replace natural methods of seed dispersal; chemical pesticides replace natural controls on populations of weeds, insects, and pathogens; and genetic manipulation replaces natural processes of plant evolution and selection. Even decomposition is altered since plant growth is harvested and soil fertility maintained not through biologically mediated nutrient recycling but with fertilizers (Altieri, 1995).

Another important way in which agriculture affects biodiversity is through the externalities associated with the intensive agrochemical and mechanical technology used to boost crop production. In the United States about 17.8 million tons of fertilizers are used in grain production systems, and about 500 million pounds of pesticides are applied annually to farmlands. Although these inputs have boosted crop yields, their undesirable environmental effects are undermining the sustainability of agriculture. Environmental (including the loss of key biodiversity elements such as pollinators and natural enemies) and social costs associated with pesticide use today are crudely estimated to reach more than $850 million per year in the U.S. (Pimentel et al., 1980). About 1.5 million hectares worldwide and about 27 percent of the irrigated land in the United States is damaged by salinization due to excessive or improper irrigation. Because of lack of rotations and insufficient vegetation cover, soil erosion levels average 185 tons/ha per year in U.S. croplands, well above the acceptable threshold. It is estimated that soil degradation has reduced crop productivity by around 13 percent (Brown and Young, 1990).

Nowhere are the consequences of biodiversity reduction more evident than in the realm of agricultural pest management. The instability of agroecosystems becomes manifest as the worsening of most insect-pest problems is increasingly linked to the expansion of crop monocultures at the expense of the natural vegetation, thereby decreasing local habitat diversity (Altieri and Letourneau, 1982; Flint and Roberts, 1988). Plant communities that are modified to meet the special needs of humans become subject to heavy pest damage, and generally the more intensely such communities are modified, the more abundant and serious the pests. The inherent self-regulation characteristics of natural communities are lost when humans modify such communities by breaking the fragile thread of community interactions (Turnbull, 1969). This breakdown can be repaired by restoring the elements of community homeostasis through the addition or enhancement of biodiversity.

In this book, we explore practical steps to break the tendency for monoculture crop production and thus reduce ecological vulnerability by restoring agricultural biodiversity at the field and landscape level. The most obvious advantage of diversification is a reduced risk of total crop failure due to invasions by unwanted species and subsequent pest infestations.

TRADITIONAL AGROECOSYSTEMS AS MODELS OF BIODIVERSE FARMS

Not all forms of agriculture lead to the extreme simplification of biodiversity. A salient feature of traditional farming systems managed by small farmers in the third world is their degree of plant diversity in the form of polycultures and/or agroforestry patterns. In fact, the species-rich quality of all biotic components of traditional agroecosystems is comparable with that of many natural ecosystems. These systems offer a means of promoting diversity of diet and income, stability of production, minimization of risk, reduced insect and disease incidence, efficient use of labor, intensification of production with limited resources, and maximization of returns under low levels of technology. Traditional, multiple-cropping systems are estimated to still provide as much as 15 to 20 percent of the world's food supplies. In Latin America, farmers grow 70 to 90 percent of their beans in combination with maize, potatoes, and other crops. Maize is intercropped in 60 percent of the region's maize-growing area (Francis, 1986).

Traditional cropping systems are also genetically diverse, containing numerous varieties of domesticated crop species as well as their wild relatives. In the Andes, farmers cultivate as many as 50 potato varieties in their fields. Maintaining genetic diversity appears to be of even greater importance as land becomes more marginal and hence farming more risky. In Peru, for example, the number of potato varieties cultivated increases with the altitude of the land farmed. Genetic diversity confers at least partial resistance to diseases that are specific to particular strains of crops and allows farmers to exploit different soil types and microclimates for a variety of nutritional and other uses (Brush, 1982).

On the other hand, traditional agroforestry systems throughout the tropics commonly contain well over 100 annual and perennial plant species per field, species used for construction materials, firewood, tools, medicine, livestock feed, and human food. In these systems, besides providing useful products, trees minimize nutrient leaching and soil erosion and restore key nutrients by pumping them from the lower soil strata (Nair, 1993). Examples include the home gardens of the Huastec Indians in Mexico and the agroforestry systems of the Amazonian Kayapo and Bora Indians (Toledo et al., 1985).

Intercropping, agroforestry, shifting cultivation, and other traditional farming methods mimic natural ecological processes, and their sustainability lies in the ecological models they follow. This use of natural analogies suggests principles for the design of agricultural systems that make effective use of sunlight, soil nutrients, rainfall, and biological resources. Several scientists now recognize how traditional farming systems can serve as models of efficiency as these systems incorporate careful management of soil, water, nutrients, and biological resources.

In developing countries, biodiversity can be used to help the great mass of resource-poor farmers, mostly confined to marginal soils, hillsides, and rain-fed areas, to achieve year-round food self-sufficiency, reduce their reliance on scarce and expensive agricultural chemical inputs, and develop production systems that rebuild the productive capacities of their small holdings (Altieri, 1987). The objective is to assist farmers in developing sustainable farming systems that satisfy food self-sufficiency, as well as stabilize production by avoiding soil erosion (Beets, 1990). Technically, the approach consists of devising multiple-use farming systems emphasizing soil and crop protection and achieving soil-fertility improvement and crop protection through the integration of trees, animals, and crops (Figure 1.1).

Examples of grassroots rural development programs in Latin America analyzed by Altieri (1991d, 1999) suggest that the maintenance and/or enhancement of biodiversity in traditional agroecosystems represents a strategy that ensures diverse diets and income sources, stable production, minimum risk, intensive production with limited resources, and maximum returns under low levels of technology. In these systems, the complementarity of agricultural enterprises reduces the need for outside input. The correct spatial and temporal assemblage of crops, trees, animals, soil, and so forth enhances the interactions that sponsor yields dependent on internal sources and recycling of nutrients and organic matter and on trophic relationships among plants, insects, or pathogens, which enhance biological pest control (Altieri and Nicholls, 2000).

Since traditional farmers generally have a profound knowledge of biodiversity, their knowledge and environmental perceptions should

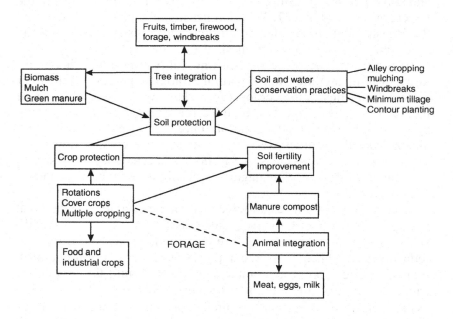

FIGURE 1.1. Assemblage of a diversified agroecosystem resulting in enhanced soil protection, soil fertility, and biological crop protection (after Altieri, 1987).

be integrated into schemes of agricultural innovation that attempt to link resource conservation and rural development (Altieri and Hecht, 1991). For a resource conservation strategy compatible with a diversified production strategy to succeed among small farmers, the process must be linked to rural development efforts that give equal importance to local resource conservation and food self-sufficiency and/or participation in local markets. Any attempt at soil, forest, or crop genetic conservation must struggle both to preserve the diversity of the agroecosystems in which these resources occur and to protect the local cultures that nurture them. Cultural diversity is as crucial as biological diversity.

THE ECOLOGICAL ROLE OF BIODIVERSITY

In addition to producing valuable plants and animals, biodiversity performs many ecological services. In natural ecosystems, the vegetative cover of a forest or grassland prevents soil erosion, replenishes groundwater, and controls flooding by enhancing infiltration and reducing water runoff. Natural habitats also contain wild populations of domesticated plants and animals, and these populations contain useful genes that are often absent in the domesticated gene pool. The entire range of our domestic crops is derived from wild species that have been modified through domestication, selective breeding, and hybridization. Most remaining world centers of diversity contain populations of variable and adaptable landraces, as well as wild and weedy relatives of crops (Harlan, 1975). Many traditionally managed farming systems in the Third World constitute in situ repositories of native crop diversity (Altieri and Hecht, 1991). There is great concern today about genetic erosion of crops in areas where small farmers are pushed by agricultural modernization to adopt new or modified varieties at the expense of traditional ones.

In agricultural systems, biodiversity performs ecosystem services beyond production of food, fiber, fuel, and income. Examples include recycling of nutrients, control of local microclimate, regulation of local hydrological processes, regulation of the abundance of undesirable organisms, and detoxification of noxious chemicals. These renewal processes and ecosystem services are largely biological; therefore, their persistence depends upon maintenance of biological diversity.

When these natural services are lost due to biological simplification, the economic and environmental costs can be quite significant. Economically, agricultural burdens include the need to supply crops with costly external inputs, since agroecosystems deprived of basic regulatory functional components lack the capacity to sponsor their own soil fertility and pest regulation. Often the costs also involve a reduction in the quality of life due to decreased soil, water, and food quality when pesticide, nitrate, or other type of contamination occurs.

Clearly, the fates of agriculture and biodiversity are intertwined. It is possible to intensify agriculture in a sustainable manner in order to secure some of the remaining natural habitats, thus ensuring the provision of environmental services to agriculture. Agroecological forms of intensification can also enhance the conservation and use of agrobiodiversity, which can lead to better use of natural resources and agroecosystem stability (Gliessman, 1999).

THE NATURE OF BIODIVERSITY IN AGROECOSYSTEMS

Biodiversity in agroecosystems can be as varied as the crops, weeds, arthropods, or microorganisms involved or the geographical location and climatic, edaphic, human, and socioeconomic factors. In general, the degree of biodiversity in agroecosystems depends on four main characteristics of the agroecosystem (Southwood and Way, 1970):

- The diversity of vegetation within and around the agroecosystem
- The permanence of the various crops within the agroecosystem
- The intensity of management
- The extent of the isolation of the agroecosystem from natural vegetation

The biodiversity components of agroecosystems can be classified in relation to the roles they play in the functioning of cropping systems. According to this, agricultural biodiversity can be grouped as follows (Swift and Anderson, 1993):

- *Productive biota:* crops, trees, and animals chosen by farmers that play a determining role in the diversity and complexity of the agroecosystem

- *Resource biota:* organisms that contribute to productivity through pollination, biological control, decomposition, etc.
- *Destructive biota:* weeds, insect pests, microbial pathogens, etc., that farmers aim at reducing through cultural management

Two distinct components of biodiversity can be recognized in agroecosystems (Vandermeer and Perfecto, 1995). The first component, *planned biodiversity,* includes the crops and livestock purposely included in the agroecosystem by the farmer, which will vary depending on the management inputs and the spatial and temporal arrangements of crops. The second component, *associated biodiversity,* includes all soil flora and fauna, herbivores, carnivores, decomposers, etc., that colonize the agroecosystem from surrounding environments and that will thrive in the agroecosystem depending on its management and structure. The relationship of both types of biodiversity components is illustrated in Figure 1.2. Planned biodiversity has a direct function, as illustrated by the bold arrow connecting the planned

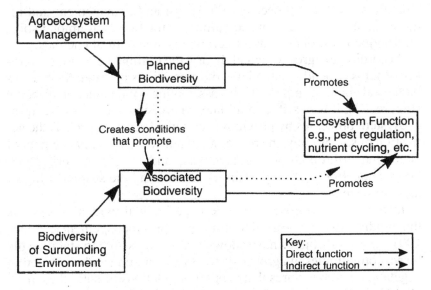

FIGURE 1.2. The relationship between planned biodiversity (that which the farmer determines based on management of the agroecosystems) and associated biodiversity and how the two promote ecosystem function (modified from Vandermeer and Perfecto, 1995).

biodiversity box with the ecosystem function box. Associated biodiversity also has a function, but it is mediated through planned biodiversity. Thus, planned biodiversity also has an indirect function, illustrated by the dotted arrow in the figure, which is realized through its influence on the associated biodiversity. For example, the trees in an agroforestry system create shade, which makes it possible to grow sun-intolerant crops. So, the direct function of this second species (the trees) is to create shade. Yet along with the trees might come wasps that seek out the nectar in the tree's flowers. These wasps may in turn be the natural parasitoids of pests that normally attack crops. The wasps are part of the associated biodiversity. The trees, then, create shade (direct function) and attract wasps (indirect function) (Vandermeer and Perfecto, 1995).

Complementary interactions between the various biotic components can also be of a multiple nature. Some of these interactions can be used to induce positive and direct effects on the biological control of specific crop pests, soil-fertility regeneration and/or enhancement, and soil conservation. The exploitation of these interactions in real situations involves agroecosystem design and management and requires an understanding of the numerous relationships among soils, microorganisms, plants, insect herbivores, and natural enemies.

According to agroecological theory, the optimal behavior of agroecosystems depends on the level of interactions between the various biotic and abiotic components. By assembling a functional biodiversity (Figure 1.3), it is possible to initiate synergisms that subsidize agroecosystem processes by providing ecological services such as the activation of soil biology, the recycling of nutrients, the enhancement of beneficial arthropods and antagonists, and so on (Altieri, 1995; Gliessman, 1999), all important in determining the sustainability of agroecosystems.

In modern agroecosystems, the experimental evidence suggests that biodiversity can be used for improved pest management (Andow, 1991a). Several studies have shown that it is possible to stabilize the insect communities of agroecosystems by designing and constructing vegetational architectures that support populations of natural enemies or have direct deterrent effects on pest herbivores.

The key is to identify the type of biodiversity that is desirable to maintain and/or enhance in order to carry out ecological services, and

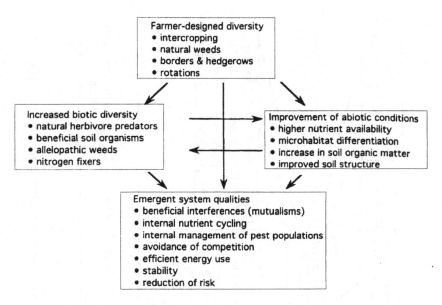

FIGURE 1.3. System dynamics in diverse agroecosystems (after Gliessman, 1999).

then to determine the best practices that will encourage the desired biodiversity components (Figure 1.4). There are many agricultural practices and designs that have the potential to enhance functional biodiversity and others that negatively affect it. The idea is to apply the best management practices in order to enhance or regenerate the kind of biodiversity that can subsidize the sustainability of agro-ecosystems by providing ecological services such as biological pest control, nutrient cycling, water and soil conservation, etc. The role of agroecologists should be to encourage those agricultural practices that increase the abundance and diversity of above- and belowground organisms, which in turn provide key ecological services to agro-ecosystems (Figure 1.5).

Thus, a key strategy of agroecology is to exploit the complementarity and synergy that result from the various combinations of crops, trees, and animals in agroecosystems that feature spatial and temporal arrangements such as polycultures, agroforestry systems, and crop-livestock mixtures. In real situations, the exploitation of

14

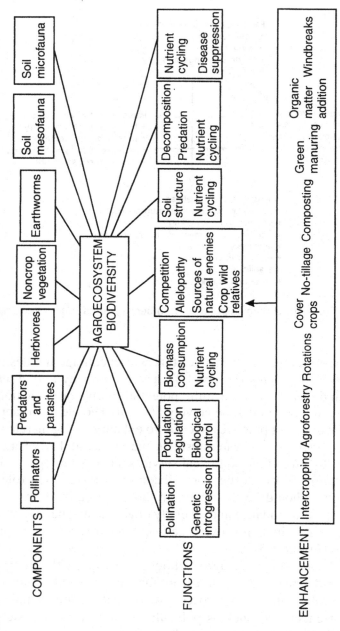

FIGURE 1.4. The components, functions, and enhancement strategies of biodiversity in agroecosystems (after Altieri, 1991a).

these interactions involves agroecosystem design and management and requires an understanding of the numerous relationships among soils, microorganisms, plants, insect herbivores, and natural enemies. This book analyzes options for agroecosystem design in detail.

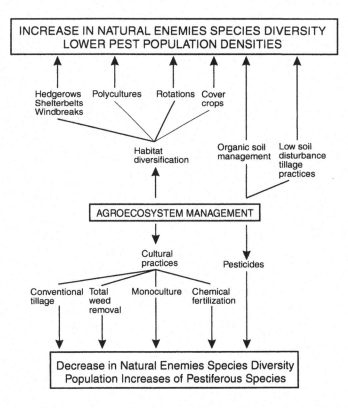

FIGURE 1.5. The effects of agroecosystem management and associated cultural practices on the biodiversity of natural enemies and the abundance of insect pests (after Altieri and Nicholls, 2000).

Chapter 2

Agroecology and Pest Management

THE NATURE OF AGRICULTURAL HABITATS AND ITS RELATION TO PEST BUILDUP

Each region has a unique set of agroecosystems that result from local climate, topography, soil, economic relations, social structure, and history. Each region contains a hierarchy of systems (Figure 2.1) in which the regional system is a complex of land-utilization units with farming subsystems and cropping subsystems which produce and transform primary products, involving a large service sector including urban centers (Hart, 1980). The agroecosystems of a region often include both commercial and local-use agricultural systems that rely on technology to different extents depending on the availability of land, capital, and labor. Some technologies in modern systems aim at efficient land use (reliance on biochemical inputs); others reduce labor (mechanical inputs). In contrast, resource-poor farmers usually adopt low-input technology and labor-intensive practices that optimize production efficiency and recycle scarce resources (Matteson, Altieri, and Gagne, 1984).

Although each farm is unique, farms can be classified by type of agriculture or agroecosystem. Functional grouping is essential for devising appropriate management strategies. Five criteria can be used to classify agroecosystems in a region:

1. The types of crop and livestock
2. The methods used to grow the crops and produce the stock
3. The relative intensity of use of labor, capital, and organization, and the resulting output of product

4. The disposal of the products for consumption (whether used for subsistence or supplement on the farm or sold for cash or other goods)
5. The structures used to facilitate farming operations (Norman, 1979)

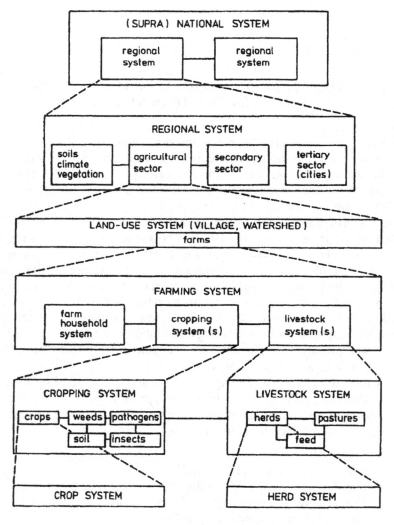

FIGURE 2.1. Agriculture as a hierarchy of systems (after Hart, 1980).

Based on these criteria, it is possible to recognize seven main types of agricultural systems in the world (Grigg, 1974):

1. Shifting cultivation systems
2. Semipermanent rain-fed cropping systems
3. Permanent rain-fed cropping systems
4. Arable irrigation systems
5. Perennial crop systems
6. Grazing systems
7. Systems with regulated ley farming (alternating arable cropping and sown pasture)

Systems 4 and 5 have evolved into habitats that are much simpler in form and poorer in species than the others, which can be considered more diversified, permanent, and less disturbed. Within the range of world agricultural systems, traditional polycultures require less energy and external inputs than modern orchards, field crops, and vegetable cropping systems to achieve a similar level of desired stability (Figure 2.2). This greater stability apparently results from certain ecological and management attributes inherent to polycultural

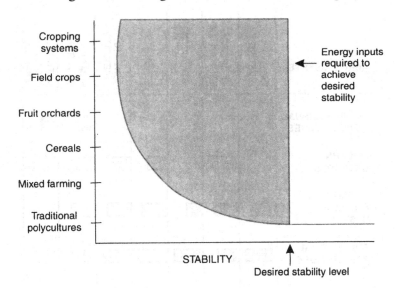

FIGURE 2.2. Energy requirements to sustain a desired level of production stability in a range of farming systems (after Altieri, 1987).

cropping systems. Modern systems require more radical modifications of their structure to approach a more diversified, less disturbed state.

Across the world, agroecosystems differ in age, diversity, structure, and management. In fact, there is great variability in basic ecological and agronomic patterns among the various dominant agroecosystems (Figure 2.3). In general, agroecosystems that are more diverse, more permanent, isolated, and managed with low-input technology (i.e., agroforestry systems, traditional polycultures) take fuller advantage of work usually done by ecological processes associated with higher biodiversity than highly simplified, input-driven, and disturbed systems (i.e., modern vegetable monocultures and orchards).

All agroecosystems are dynamic and subjected to different levels of management so that the crop arrangements in time and space are continually changing in the face of biological, cultural, socioeconomic, and environmental factors. Such landscape variations determine the degree of spatial and temporal heterogeneity characteristic of agricultural regions, which may or may not benefit the pest protection of particular agroecosystems. Thus, one of the main challenges

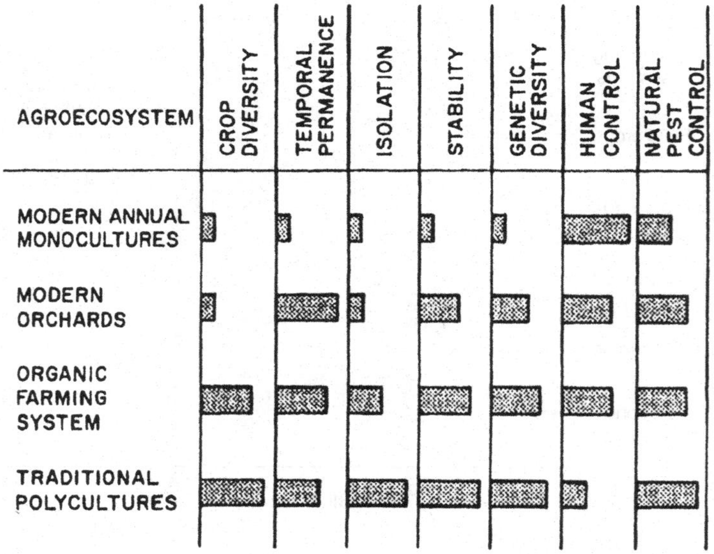

AGROECOSYSTEM	CROP DIVERSITY	TEMPORAL PERMANENCE	ISOLATION	STABILITY	GENETIC DIVERSITY	HUMAN CONTROL	NATURAL PEST CONTROL
MODERN ANNUAL MONOCULTURES	▪	▪	▪	▪	▪	▬▬	▬
MODERN ORCHARDS	▪	▬▬	▪	▬	▬	▬	▬
ORGANIC FARMING SYSTEM	▬	▬	▬	▬	▬	▬	▬▬
TRADITIONAL POLYCULTURES	▬▬	▬	▬▬	▬▬	▬▬	▪	▬▬

FIGURE 2.3. Ecological patterns of contrasting agroecosystems (after Altieri, 1987). Longer bars indicate a higher degree of the characteristic.

facing agoecologists today is identifying the types of heterogeneity (either at the field or regional level) that will yield desirable agricultural results (i.e., pest regulation), given the unique environment and entomofauna of each area. This challenge can be met only by further analyzing the relationship between vegetation diversification and the population dynamics of herbivore species in light of the diversity and complexity of site-specific agricultural systems. A hypothetical pattern in pest regulation according to agroecosystem temporal and spatial diversity is depicted in Figure 2.4. According to this "increasing probability for pest buildup" gradient, agroecosystems on the left side of the gradient are more biodiverse and tend to be more amenable to manipulation since polycultures already contain many of the key factors required by natural enemies. There are, however, habitat manipulations that can introduce appropriate diversity into the important (but biodiversity impoverished) grain, vegetable, and row crop systems lying in the right half of Figure 2.4.

Although herbivores vary widely in their response to crop distribution, abundance, and dispersion, the majority of agroecological studies show that structural (i.e., spatial and temporal crop arrangement) and management (i.e., crop diversity, input levels, etc.) attributes of agroecosystems influence herbivore dynamics. Several of these attributes are related to biodiversity, and most are amenable to management (i.e., crop sequences and associations, weed diversity, genetic diversity, etc.).

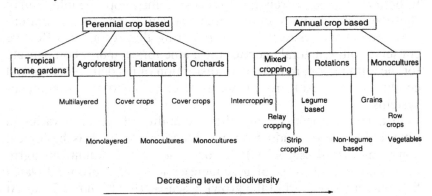

FIGURE 2.4. A classification of dominant agricultural agroecosystems on a gradient of diversity and vulnerability to pest outbreak (after Altieri, 1991c).

Crop temporal and spatial patterns in agroecosystems throughout the world vary tremendously. There has been little analysis and discussion of whether temporal and spatial patterns that characterize crop phenology in agroecosystems influence the potential success of biological control through conservation. Crop spatial and temporal patterns determine the extent and persistence of crop plants and, thus, the availability of key resources associated with the crop (Kareiva, 1983; Van Emden, 1990). The spatial and temporal availability of those crop-associated resources (and those in surrounding or adjacent unmanaged habitats or in managed refuges) also may be critical determinants of whether, when, or how they respond to herbivores or other resources provided by crops (or other plants). Indeed, in simulation studies, Corbett and Plant (1993) have noted that the timing of the availability of interplanted (refuge) vegetation relative to the germination of crop plants may determine if the refuge is likely to act as a source of natural enemies or as a sink (i.e., taking natural enemies away from crops).

Agroecosystems are managed habitats with concentrations of perennial crops, annual crops, or both. The crop plant's life cycle, to a great extent, dictates the nature of the habitat (i.e., its structure and texture, longevity, and the composition and complexity of its fauna and flora). The permanence of a crop determines the intensity and complexity of the interactions that unfold in a given agroecosystem. In both annual and perennial agroecosystems, crop phenology may cause asynchrony between resource availability and the natural enemy (predator or parasite) stage requiring that resource. Effective conservation of natural enemies must ameliorate and/or compensate for the elimination, reduction, or disruption of needed resources and conditions that result from patterns of crop phenology in agroecosystems.

In annual agroecosystems, the availability of the crop varies in time and space depending on agroeconomic as well as biological constraints (Barbosa, 1998). At one end of a gradient, an agroecosystem may consist of a sequence of a single crop cultivated throughout most or all of the growing season (see Figure 2.5, A). At the other end of the gradient, an agroecosystem may be characterized by a sequence of plantings and harvests of different crops (see C). A third point in this hypothetical gradient is represented by agroeco-

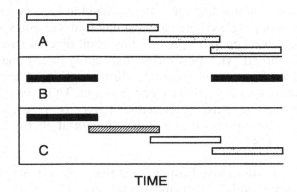

TIME

FIGURE 2.5. Hypothetical sequence of crop plants. Bars with different patterns represent different crops or crop cultivars (after Barbosa, 1998).

systems in which a given crop may occur discontinuously at two different times during a season (see B).

Implementation of biological control tactics will vary among crop sequences. In sequence 1A, resource diversity for natural enemies is low, and the monoculture ensures completion of pest life cycles. Sequence 1C provides more diversity of resources for natural enemies and breaks the pest life cycles more effectively.

In a discontinuous cropping system implementation of conservation biological control during one or both crop phases is likely to have little impact without a thorough plan for the conservation of natural enemies during the interval between the first and the second crop plantings. Natural enemies, particularly monophagous species, must survive when the crop and hosts are not present. Alternatively, the conservation of natural enemies may involve reliance on or manipulations of unmanaged habitats (or managed refuges) in the landscape of the agroecosystem to compensate for the discontinuous crop pattern.

CROP DIVERSIFICATION AND BIOLOGICAL CONTROL

Crop monocultures are difficult environments in which to induce efficient biological pest control because these systems lack adequate resources for effective performance of natural enemies and because disturbing cultural practices are often utilized in such systems. More diversified cropping systems already contain certain specific resources for natural enemies provided by plant diversity and are usually not disturbed with pesticides (especially when managed by resource-poor farmers who cannot afford high-input technology). They are also more amenable to manipulation. Thus, by replacing or adding diversity to existing systems, it may be possible to exert changes in habitat diversity that enhance natural enemy abundance and effectiveness by

1. providing alternative hosts/prey at times of pest-host scarcity;
2. providing food (pollen and nectar) for adult parasitoids and predators;
3. providing refuges for overwintering, nesting, and so on; and
4. maintaining acceptable populations of the pest over extended periods to ensure continued survival of beneficial insects (van den Bosch and Telford, 1964; Altieri and Letourneau, 1982; Powell, 1986).

The specific resulting effect or the strategy to use will depend on the species of herbivores and associated natural enemies, as well as on properties of the vegetation, the physiological condition of the crop, or the nature of the direct effects of particular plant species (Letourneau, 1987). In addition, the success of enhancement measures can be influenced by the scale upon which they are implemented (i.e., field scale, farming unit, or region) since field size, within-field and surrounding vegetation composition, and level of field isolation (i.e., distance from source of colonizers) will all affect immigration rates, emigration rates, and the effective tenure time of a particular natural enemy in a crop field. Whatever diversity enhancement strategy is used, it must be based on a thorough knowledge of the natural enemies' ecological requirements.

Perhaps one of the best strategies to increase effectiveness of predators and parasitoids is the manipulation of nontarget food resources (i.e., alternate hosts/prey and pollen/nectar) (Rabb, Stinner, and van den Bosch, 1976). Here it is important that not only the density of the nontarget resource be sufficiently high to influence enemy populations but also that the spatial distribution and temporal dispersion of the resource be adequate. Proper manipulation of the nontarget resource should result in the enemies colonizing the habitat earlier in the season than the pest and frequently encountering an evenly distributed resource in the field, thus increasing the probability that the enemy will remain in the habitat and reproduce (Andow and Risch, 1985). Certain polycultural arrangements increase and others reduce the spatial heterogeneity of specific food resources; thus, particular species of natural enemies may be more or less abundant in a specific polyculture. These effects and responses can be determined only experimentally across a whole range of agroecosystems. The task is indeed overwhelming since enhancement techniques must necessarily be site specific.

The literature is full of examples of experiments documenting that diversification of cropping systems often leads to reduced herbivore populations. The studies suggest that the more diverse the agroecosystem and the longer this diversity remains undisturbed, the more internal links develop to promote greater insect stability (Way, 1977). It is clear, however, that the stability of the insect community depends not only on its trophic diversity but also on the actual density-dependence nature of the trophic levels (Southwood and Way, 1970). In other words, stability will depend on the precision of the response of any particular trophic link to an increase in the population at a lower level.

Although most experiments have moved forward on documenting insect population trends in single versus complex crop habitats, a few have concentrated on elucidating the nature and dynamics of the trophic relationships between plants and herbivores and herbivores and their natural enemies in diversified agroecosystems. Several lines of study have developed.

- *Crop-weed-insect interaction studies:* Evidence indicates that weeds influence the diversity and abundance of insect herbivores and associated natural enemies in crop systems. Certain flowering weeds (mostly Umbelliferae, Leguminosae, and Compositae) play an important ecological role by harboring and supporting a complex of beneficial arthropods that aid in suppressing pest populations (Altieri, Schoonhoven, and Doll, 1977; Altieri and Whitcomb, 1979b, 1980).
- *Insect dynamics in annual polycultures:* Overwhelming evidence suggests that polycultures support a lower herbivore load

than monocultures do. One factor explaining this trend is that relatively more stable natural-enemy populations can persist in polycultures due to the more continuous availability of food sources and microhabitats (Letourneau and Altieri, 1983; Helenius, 1989). The other possibility is that specialized herbivores are more likely to find and remain on pure crop stands that provide concentrated resources and monotonous physical conditions (Tahvanainen and Root, 1972).

- *Herbivores in complex perennial crop systems:* Most of these studies have explored the effects of the manipulation of groundcover vegetation on insect pests and associated enemies. The data indicate that orchards with rich floral undergrowth exhibit a lower incidence of insect pests than clean-cultivated orchards, mainly because of an increased abundance and efficiency of predators and parasitoids (Altieri and Schmidt, 1985). In some cases, groundcover directly affects herbivore species that discriminate among trees with and without cover beneath.

- *The effects of adjacent vegetation:* These studies have documented the dynamics of colonizing insect pests that invade crop fields from edge vegetation, especially when the vegetation is botanically related to the crop. A number of studies document the importance of adjoining wild vegetation in providing alternate food and habitat to natural enemies that move into nearby crops (Van Emden, 1965b; Wainhouse and Coaker, 1981; Altieri and Schmidt, 1986a).

The available literature suggests that the design of vegetation-management strategies must include knowledge and consideration of (1) crop arrangement in time and space, (2) the composition and abundance of noncrop vegetation within and around fields, (3) the soil type, (4) the surrounding environment, and (5) the type and intensity of management. The response of insect populations to environmental manipulations depends upon their degree of association with one or more of the vegetational components of the system. Extension of the cropping period or planning temporal or spatial cropping sequences may allow naturally occurring biological control agents to attain higher population levels on alternate hosts or prey and to persist in the agricultural environment throughout the year.

Since farming systems in a region are managed over a range of energy inputs, levels of crop diversity, and successional stages, variations in insect dynamics are likely to occur and may be difficult to predict. However, based on current ecological and agronomic theory, low pest potentials may be expected in agroecosystems that exhibit the following characteristics:

1. High crop diversity through mixtures in time and space (Cromartie, 1981; Altieri and Letourneau, 1982; Risch, Andow, and Altieri, 1983; but see also Andow and Risch, 1985; Nafus and Schreiner, 1986)
2. Discontinuity of monoculture in time through rotations, use of short-maturing varieties, use of crop-free or preferred host-free periods, etc. (Stern, 1981; Lashomb and Ng, 1984)
3. Small, scattered fields creating a structural mosaic of adjoining crops and uncultivated land which potentially provide shelter and alternative food for natural enemies (Van Emden, 1965a; Altieri and Letourneau, 1982) (Pests also may proliferate in these environments depending on plant species composition [Altieri and Letourneau, 1982; Slosser et al., 1984; Collins and Johnson, 1985; Levine, 1985; Lasack and Pedigo, 1986]. However, the presence of low levels of pest populations and/or alternate hosts may be necessary to maintain natural enemies in the area.)
4. Farms with a dominant perennial crop component. (Orchards are considered to be semipermanent ecosystems and more stable than annual crop systems. Since orchards suffer less disturbance and are characterized by greater structural diversity, possibilities for the establishment of biological control agents are generally higher, especially if floral undergrowth diversity is encouraged [Huffaker and Messenger, 1976; Altieri and Schmidt, 1985].)
5. High crop densities or presence of tolerable levels of weed background (Shahjahan and Streams, 1973; Altieri, Schoonhoven, and Doll, 1977; Andow, 1983b; Mayse, 1983; Buschman, Pitre, and Hodges, 1984; Ali and Reagan, 1985)
6. High genetic diversity resulting from the use of variety mixtures or several lines of the same crop (Perrin, 1977; Gould, 1986; Altieri and Schmidt, 1987)

By considering various spatial, temporal, and varietal features of cropping systems, Litsinger and Moody (1976) suggested the implications for pest suppression of various crop-management schemes (Figure 2.6). These generalizations can serve in the planning of a vegetation-management strategy in agroecosystems; however, they must take into account local variations in climate, geography, crops, local vegetation, inputs, pest complexes, etc., which might increase or decrease the potential for pest development under some vegetation-management conditions. The selection of component plant species can also be critical. Systematic studies on the "quality" of plant diversification with respect to the abundance and efficiency of natural enemies are needed. As pointed out by Southwood and Way (1970), what seems to matter is "functional" diversity and not diversity per se. This highlights the importance of recognizing that agrecosystems may not benefit from a "hit-and-miss" approach to diversification but require certain elements of diversity which, once identified, could be retained or reintroduced. Mechanistic studies to determine the underlying elements of plant mixtures which disrupt pest invasion and which favor colonization and population growth of natural enemies will allow more precise planning of cropping schemes and increase the chances of a beneficial effect beyond the current levels. It is important that changes in habitat diversity are purposely designed to obtain specific effects within the socioeconomic constraints of the agroecosystem.

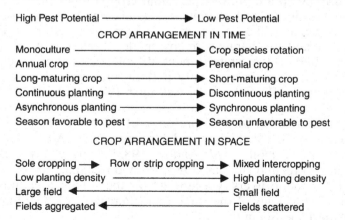

PEST POTENTIAL RELATED TO CROP MANAGEMENT

High Pest Potential ──────────────▶ Low Pest Potential

CROP ARRANGEMENT IN TIME

Monoculture ──────────────▶ Crop species rotation
Annual crop ──────────────▶ Perennial crop
Long-maturing crop ──────────▶ Short-maturing crop
Continuous planting ──────────▶ Discontinuous planting
Asynchronous planting ────────▶ Synchronous planting
Season favorable to pest ──────▶ Season unfavorable to pest

CROP ARRANGEMENT IN SPACE

Sole cropping ──▶ Row or strip cropping ──▶ Mixed intercropping
Low planting density ──────────▶ High planting density
Large field ◀────────────── Small field
Fields aggregated ◀────────── Fields scattered

FIGURE 2.6. Hypothetical trends of increased or decreased pest potential in agroecosystems depending on crop arrangement in time and/or space (after Litsinger and Moody, 1976).

Chapter 3

Plant Diversity and Insect Stability in Agroecosystems

ECOLOGICAL THEORY

Monocultures are dominated by a single plant species and, therefore, represent an extreme example of agroecosystems with low diversity. Such systems are more susceptible to weather disasters, pest or disease outbreaks, and other catastrophes. A high degree of management and external inputs is required to maintain these types of agroecosystems. In contrast, many natural ecosystems appear to be more stable and less subject to fluctuations in populations of their component organisms.

Ecosystems with higher diversity are more stable because they exhibit higher

- resistance, or an ability to avoid or withstand disturbance; and
- resilience, or an ability to recover following disturbance.

Diversity is only one measure of ecosystem complexity. The community of organisms becomes more complex when a larger number of different kinds of organisms is included, when there are more interactions among organisms, and when the strength of these interactions increases. As diversity increases, so do opportunities for coexistence and beneficial interference between species that can enhance agroecosystem sustainability. Diverse systems encourage complex food webs that entail more potential connections and interactions among members, and many alternative paths of energy and material flow through it. Thus, a more complex community is more stable, and much data supports this idea.

Nevertheless, ecologists have for years debated the assumption that increased diversity fosters stability. Critical theoretical reviews on this subject are available (Watt, 1973; Van Emden and Williams, 1974; Goodman, 1975; Murdoch, 1975), as are reviews that use agricultural examples to bolster the theory (Pimentel, 1961; Root, 1973; Dempster and Coaker, 1974; Litsinger and Moody, 1976; Perrin, 1977).

Regardless of those discussions, research has shown that mixing certain plant species with the primary host of a specialized herbivore gives a fairly consistent result (Figure 3.1): specialized species usually exhibit higher abundance in monocultures than in polycultures. In a review of 150 published investigations, Risch and colleagues (1983) found evidence to support the notion that specialized insect herbivores were less numerous in diverse systems (53 percent of 198 cases). Another comprehensive review by Andow (1991a) identified 209 published studies that deal with the effects of vegetation diversity in agroecosystems on herbivorous arthropod species. Fifty-two percent of the 287 total herbivore species examined in these studies were found to be less abundant in diversified systems than in monocultures, while only 15.3 percent (44 species) exhibited higher densities in polycultures (Table 3.1). In a more recent review of 287 cases,

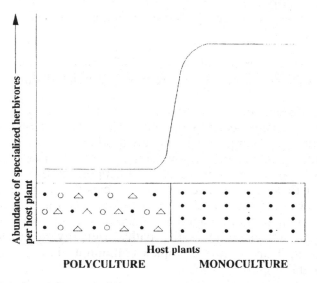

FIGURE 3.1. Consistent population trends of specialized herbivores in monocultures and polycultures where host plants are mixed with nonhost plants (after Strong, Lawton, and Southwood, 1984).

TABLE 3.1. Numbers of Arthropod Species with Particular Responses to Additive and Substitutive Polycultures [a]

Population density of arthropod species in polyculture compared to monoculture:

	Variable[b]	Higher	No Change	Lower
Herbivores	58	44	36	149
	(20.2)	(15.3)	(12.5)	(51.9)
Monophagous	42	17	31	130
	(19.1)	(7.7)	(14.1)	(59.1)
Polyphagous	16	27	5	19
	(23.9)	(40.3)	(7.5)	(28.4)
Natural enemies	33	68	17	12
	(25.6)	(52.7)	(13.2)	(9.3)
Predators	27	38	14	11
	(30.3)	(42.7)	(15.7)	(12.4)
Parasitoids	6	30	3	1
	(15.0)	(75.0)	(7.5)	(2.5)

Source: After Andow, 1991a.

[a]Percentage of total number of species is in parentheses.
[b]A variable response means that an arthropod species did not consistently have a higher or lower population density in polycultures compared to monocultures when the species response was studied several times.

Helenius (1998) found that reduction of monophagous pests was greater in perennial systems and that the reduction of polyphagous pest numbers was less in perennial than in annual systems (Table 3.2). Four main ecological hypotheses have been offered to explain lower pest-population loads in multispecies plant associations.

Associational Resistance

Ecosystems in which plant species are intermingled possess an associational resistance to herbivores in addition to the resistance of individual plant species (Root, 1975). Tahvanainen and Root (1972) suggest that, in addition to their taxonomic diversity, polycultures have a relatively complex structure, chemical environment, and associated patterns of microclimates. These factors of mixed vegetation

TABLE 3.2. Percentage of Cases with Lowered Numbers of Arthropod Herbivores in Crops of Increased Vegetational Diversity Than in Monocrops

	Monophagous Species	Polyphagous Species	All
Annual systems	53.5	33.3	48.5
Perennial systems	72.3	12.5	60.5
All	59.1	28.4	51.9

Source: From Andow, 1991a.

work synergistically to produce an "associational resistance" to pest attack. In stratified vegetation, insects may experience difficulty in locating and remaining in small, favorable spots if microclimatic conditions are highly fractionated. Thus, diversity ameliorates the herbivore pressure on the crop system as a whole.

Host-plant finding by insect pests often involves olfactory mechanisms, and host plants grown in association with unrelated plants may be an important component in the defense against herbivores, the nonhost-plant odors leading to disruption of the host-finding behavior of the insect based on odor cues. This type of protection derives from the masking effect of the nonhost-plant odors on the odors emitted by the host plants. This effect has been demonstrated with collards interplanted with tomato or tobacco on the flea beetle, *Phyllotreta cruciferae* (Root, 1973), and the diamondback moth, *Plutella xylostella* (Litsinger and Moody, 1976), and the carrot fly on carrots interplanted with onions (Uvah and Coaker, 1984). In the latter example, reduction in infestation occurred only when the onion leaves were expanding and not when the plants had started to bulb, suggesting that the masking odor emanated from young leaves only. Other aromatic herbs have been claimed, mostly by organic gardeners, to be repellent to insect pests of vegetable crops, but little experimental work has been done to substantiate these claims.

Natural Enemy Hypothesis

This proposition predicts that there will be a greater abundance and diversity of natural enemies of pest insects in polycultures than in monocultures (Root, 1973). Predators tend to be polyphagous and have broad habitat requirements, so they would be expected to encounter a greater array of alternative prey and microhabitats in a heterogeneous environment (Root, 1975). Annual crop monocultures do not provide adequate alternative sources of food (pollen, nectar, prey), shelter, and breeding and nesting sites for the effective performance of natural enemies (Rabb, Stinner, and van den Bosch, 1976).

The natural-enemy hypothesis has been stated in the following way:

1. A greater diversity of prey and microhabitats is available within complex environments. As a result, relatively stable populations of generalized predators can persist in these habitats because they can exploit the wide variety of herbivores which become available at different times or in different microhabitats (Root, 1973).
2. Specialized predators are less likely to fluctuate widely because the refuge provided by a complex environment enables their prey to escape widespread annihilation (Risch, 1981).
3. Diverse habitats offer many important requisites for adult predators and parasites, such as nectar and pollen sources, which are not available in a monoculture, reducing the probability that they will leave or become locally extinct (Risch, 1981).

According to Root's enemies hypothesis, generalist and specialist natural enemies are expected to be more abundant in polycultures and, therefore, more effectively suppress herbivore population densities in polycultures than in monocultures. Generalist predators and parasitoids should be more abundant in polycultures than monocultures because (1) they switch and feed on the greater variety of herbivores that become available in polycultures at different times during the growing season; (2) they maintain reproducing populations in polycultures although in monocultures only males of some parasitoids are produced; (3) they can utilize hosts in polycultures that they would normally not encounter and use in monocultures; (4) they can exploit the greater variety of herbivores available in different microhabitats in the polycultures; and (5) prey or hosts are more abundant or more available in polycultures (Smith and McSorley, 2000).

Specialist predator and parasitoid populations are expected to be more abundant and effective in polycultures than monocultures because prey or host refuges in polycultures enable the prey or host populations to persist, which stabilizes predator-prey and parasitoid-host interactions. In monocultures, predators and parasitoids drive their prey or host populations to extinction and become extinct themselves shortly thereafter. Prey or host populations will recolonize these monocultures and rapidly increase (Andow, 1991a).

Finally, both generalist and specialist natural enemies should be more abundant in polycultures than monocultures because more pollen and nectar resources are available at more times during the season in the complex systems (Altieri and Letourneau, 1982).

Resource Concentration

Insect populations can be influenced directly by the concentration or spatial dispersion of their food plants. There can be a direct effect of associated plant species on the ability of the insect herbivore to find and utilize its host plants. Many herbivores, particularly those with narrow host ranges, are more likely to find and remain on hosts that are growing in dense or nearly pure stands (Root, 1973), which are thus providing concentrated resources and monotonous physical conditions.

For any pest species, it is the total strength of the attractive stimuli that determines the resource concentration, and this varies with interacting factors such as the density and spatial arrangements of the host plant and the interfering effects of nonhost plants. Consequently, the lower the resource (host plant) concentration, the more difficult it will be for the insect pest to locate a host plant. Relative resource concentration also increases the probability of the pest species leaving the habitat once it has arrived; for instance, the pest may tend to fly sooner and further after landing on a nonhost plant, which can result in a higher emigration rate from polycultures than from monocultures (Andow, 1991a). Such an "emigration effect" should be evident in polycultures when the pest's trivial movement involves (1) mistakenly alighting on nonhost plants, (2) moving off nonhost plants more frequently than moving off host plants, and (3) being at risk of leaving the crop area during movement.

Using diffusion models (calculating movement coefficients and disappearance rates), Power (1987) compared movement rates of the leafhopper *Dalbulus maidis* in maize monocultures and maize intercropped with beans. The rate of movement along rows and the disappearance rate were twice as rapid in the intercrop as in the monoculture, but the rate of movement across rows was dramatically reduced in the intercrop. As would be expected from the disruptive crop, movement along the rows was more rapid when an effective block of beans was between the rows, suggesting a more rapid rate of disappearance in the intercrop. This hypothesis has also been experimentally tested by Bach (1980b) and Risch (1981).

Plant "Apparency"

Most crops are derived from early successional herbs that escaped from herbivores in space and time (Feeny, 1976). The effectiveness of natural crop-plant defenses is reduced by present agricultural methods: monocultures make crop plants "more apparent" to herbivores than were their ancestors. In agriculture, the "apparency" of a crop plant is increased by close association with related species (Feeny, 1977). Therefore, crop plants grown in monoculture are subjected to artificial conditions for which their qualitative chemical and physical defenses are inadequate. The theory of plant chemical defense developed by Feeny (1976) and Rhoades and Cates (1976) discusses the classification of plants as "apparent or predictable" and "unapparent or unpredictable" and the implications of such divisions for agricultural crops in relation to pest susceptibility.

Crop apparency can be increased or decreased either by intracrop diversity or by high-density cropping. Vegetational background among crops can have different effects on the associated insect fauna depending on the situation the pest is adapted to exploit.

Pieris rapae and *Brevicoryne brassicae* favor mainly open successional habitats and are more attracted to host plants that stand out against a bare soil; in contrast, the fruit fly dwells in dense stands and would be less attracted to open plantings of grasses and cereals (Burn, Coaker, and Jepson, 1987).

Critical Assessments of the Hypotheses

Of the four hypotheses, insect ecologists have tested more systematically the "enemy hypothesis" (predators and parasites are more effective in complex systems) and the "resource concentration hypothesis" (specialist herbivores more easily find, stay in, and reproduce in simple systems). Table 3.3 summarizes the main findings and conclusions of major reviews that explore the ecological implications of both hypotheses.

Finch and Collier (2000) argue that both hypotheses fail to produce a general theory of host-plant selection, and thus, they propose a new theory based on "appropriate/inappropriate landings." This theory is based on the fact that during the host-plant-finding phase, the searching insects land indiscriminately on leaves of host plants (appropriate landings) and nonhost plants (inappropriate landings) but avoid landing on brown surfaces, such as soil. This process seems to be governed by visual stimuli, as opposed to chemical cues, a central link within the three-link chain of events that governs host-plant selection.

In a study of eight different phytophagous species, Finch and Kienegger (1997) found that the ability of each species to find cabbage was affected adversely, though to differing degrees, when their host plants were surrounded by clover. From a crop-protection standpoint, the more host plants are concentrated in a given crop area, the greater the chance an insect has of finding a host plant. Current agricultural methods are exacerbating pest control problems, as "bare-soil" cultivation ensures that crop plants are exposed to the maximum pest-insect attack possible in any given locality.

One main criticism of the developing theories of insect-plant interactions in diverse agroecosystems has been advanced by Price and colleagues (1980). They contend that this theory has involved only the plant (first trophic level) and the herbivore (the second level) but has not seriously considered the natural enemies that prey on, or parasitize, the herbivores (third trophic level); this level should be viewed as part of a plant's battery of defenses against herbivores. This is especially relevant in multispecies crop associations because (1) the herbivore-enemy interaction of one plant species can be influenced by the presence of associated plants, and (2) the herbivore-enemy interaction of one plant species can be influenced by the presence of herbivores on associated plant species. In many cases, entomophagous

TABLE 3.3. Main Findings and Conclusions of Reviews of Enemy Hypothesis (NE) and Resource Concentration (RC) Hypothesis

Review	Main Findings	Main Conclusions
Risch et al. (1983)	150 studies, 198 herbivore species. "Diverse" systems: herbivore populations lower in 58 percent of cases, higher in 18 percent, no change in 9 percent, and variable in 20 percent.	Resource concentration hypothesis (Root, 1973) most likely explanation but *mechanisms* rarely studied, e.g., host-plant finding behavior, predation rate, etc.
Risch (1981)	Review of mechanisms.	Resource concentration hypothesis in *annual* systems, natural enemies in *perennial* systems.
Kareiva (1983)	Modeling and field studies show that simple systems can be stable. Species number and "connectance" are important in stability.	Resource concentration (RC) hypothesis supported rather than an effect via the community's trophic structure.
Russell (1989)	Reviewed nineteen studies which explicitly tested the natural enemies hypothesis (NE).	Of the 19 studies, mortality rates from predators and parasitoids in "diverse" systems were higher in 9, lower in 2, unchanged in 3, and variable in 5. When mechanisms are evaluated, natural enemy hypothesis more likely to be confirmed. This hypothesis and resource concentration hypothesis "complementary."
Baliddawa (1985)	Reviewed crop/weed systems and intercrops.	Crop/weed studies: 56 percent of pest decreases caused by natural enemies. Intercrops: 25 percent. Intercrops probably slowed colonization by pests. (Plants may limit predator movement also—see text.)
Andow (1991a)	Classified pests as monophagous or polyphagous. 254 herbivore species reviewed.	56 percent of herbivore species' populations were lower in diverse systems, compared with 66 percent for monophagous herbivores and 27 percent for polyphagous ones. Predator numbers were higher in 48 percent of diverse habitats studied, as were 81 percent of parasitoids'. Concluded that both hypotheses applied but RC hypothesis most important.
Cromartie (1991)	Reviewed data from annual row crops and from perennial crops.	RC hypothesis most likely for annual crops, with NE hypothesis most likely for perennial (and some annual) crops.

37

insects are directly attracted to particular plants, even in the absence
of host or prey, or by chemicals released by the herbivore's host plant
or other associated plants. Interactions can be very complex and typically involve different trophic levels and several species of plants,
herbivores, and natural enemies, as illustrated by Price and colleagues (1980) (Figure 3.2).

An additional criticism raised by Vandermeer (1981) is that none
of the hypotheses includes what Vandermeer calls the trap-cropping
effect. The idea is that the presence of a second crop in the vicinity of
a principal crop attracts a pest which would otherwise normally attack the principle crop. Corn, when planted in strips in cotton fields,
reportedly may attract the cotton bollworm away from the cotton

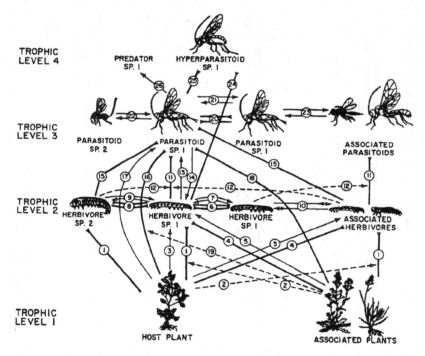

FIGURE 3.2. Interactions in a community of four trophic levels involving
semiochemicals. Arrows are placed against the responding organism. Thick
solid lines and solid arrows illustrate attraction to a stimulus (e.g., 1, 4, 11, 24).
Thin solid lines and open arrows illustrate repulsion (e.g., 3, 13, 17, 26). Thin
dashed lines show indirect effects such as interference with another response
(e.g., 2, 12, 19) (after Price et al., 1980).

(Lincoln and Isley, 1947). Sorghum may act as an effective trap for the stem borer *Chilo partellus* in India. Trap crops have been used to control jassids in cotton (Ali and Karim, 1989). In Central America, Rosset and colleagues (1985) found that the attack by the armyworm *(Spodoptera sunia)* totally destroyed a monoculture of tomatoes, although an intercrop of tomatoes and beans was effective in reducing the attack to virtually zero. It was clear that the caterpillars were being attracted to the beans in the intercrop that acted as a trap crop. More specific examples of trap crops that have proven effective to attract insect pests away from target plants in a variety of cropping systems are presented in Table 3.4. A remarkable example of trap cropping from Canada concerns the use of sterile bromegrass as a trap plant for the wheat stem sawfly *Cephus cinctus*. The bromegrass traps a large proportion of the incoming sawflies. They lay their eggs there, and the larvae bore into the stems. The elegant feature of this system is that the larvae die in the brome stems not before pupating but after larval parasitoids have emerged. No control of the insect on the trap plants is therefore necessary. Moreover, the grass acts as a filter which converts pests into beneficial biomass (Van Emden and Dabrowski, 1997).

According to Vandermeer (1989), trap crops act preferentially to attract generalist herbivores in such a way that the plant to be protected is not as likely to be directly attacked. He proposes some general models to optimize the level of trap-crop concentration to increase the trap's ability to draw pests away from the crop (local-attraction probability) as opposed to attracting pests from afar (regional-attraction probability) (Figure 3.3).

THEORY DILEMMAS

Sheehan (1986) and Russell (1989) have questioned the universal validity of these theories that explain the effects of agroecosystem diversification on searching behavior and success of arthropod natural enemies and claim that such interactions are still poorly understood. None of the proposed hypotheses really include all of the mechanisms that are known to operate in the general area of diversity and pest attack (Vandermeer, 1989). Fifty percent of the eighteen studies reviewed by Russell (1989) found higher herbivore mortality rates

TABLE 3.4. Examples of Trap Crop Systems Successfully Applied in Agricultural Practice

Controlled Pests	Main Crop	Trap Crop	Location
Lygus bugs (Lygus hesperus, Lygus elisus)	cotton	alfalfa	California
Cotton boll weevil (Anthonomus grandis)	cotton cotton	cotton cotton	United States Nicaragua
Stinkbugs (Nezara viridula Euschistus spp., Acrosternum hilare, Piezodorus guildinii)	soybeans soybeans soybeans	soybeans soybeans soybeans cowpea	United States Brazil Nigeria
Mexican bean beetle (Epilachna varivestis)	soybeans	snap beans	United States
Bean leaf beetle (Cerotoma trifurcata)	soybeans	soybeans	United States
Colorado potato beetle (Leptinotarsa decemlineata)	potato potato	potato potato	USSR Bulgaria
Blossom beetle (Meligethes aeneus)	rape cauliflower	rape marigold	Finland
Pine shoot beetle (Tomicus piniperda)	pine trees	pine logs	Great Britain
Spruce bark beetle (Ips typographus)	spruce	spruce trees logs	Europe

Source: After Hokkanen, 1991.

from predation or parasitism in diverse systems, but according to Russell, a lack of adequate control in all but one study prevented researchers from concluding that the difference in mortality was what really reduced the number of herbivores in complex systems. He further argues that the enemies hypothesis and the resource concentration hypothesis act as complementary mechanisms in reducing numbers of herbivores in polycultures, and both should be enhanced simultaneously to achieve maximum control.

According to Sheehan (1986), the enemies hypothesis is simplistic in several respects. Victim location by generalist enemies may be hin-

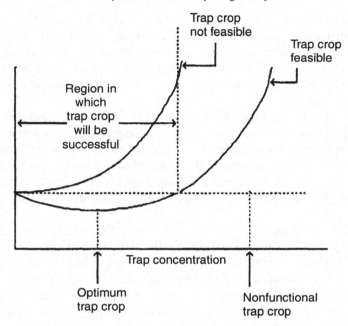

FIGURE 3.3. Ratio of general to local attraction probabilities as a function of trap concentration, illustrating the region of trap-crop feasibility (after Vandermeer, 1989).

dered by increased plant density or patchiness in diverse agricultural systems. In fact, crop diversification may reduce enemy searching efficiency and destabilize predator-prey interactions. Specialist enemies, often important in biological control programs, may be particularly sensitive to vegetation texture. Pest control by specialist enemies may be more effective in less diverse agroecosystems if concentration of host plants increases attraction or retention of these enemies. Thus, it is possible that certain specialist enemies may not necessarily respond to habitat diversification in the same way as generalists.

Sheehan (1986) suggested that specialist parasitoids might be less abundant in polycultures than monocultures because (1) chemical cues used in host finding will be disrupted and the parasitoids will be less able to find hosts to parasitize and feed upon in polycultures, and (2) the indistinct boundary at the edges of polycultures will be hard to recognize and they will be more likely to leave polycultural habitats than monocultures. In addition, Andow and Prokrym (1990) showed that structural complexity, or the connectedness of the surface on which the parasitoid searches, can strongly influence parasitoid host-finding rates. An implication of their study is that structurally complex polycultures would have less parasitism than structurally simple monocultures.

Factors that increase immigration to and decrease emigration from host-plant areas by specialist enemies (e.g., large patch size, close plant spacing, the presence of specific chemical or visual stimuli, and lower chemical or structural diversity of associated vegetation) may cause those enemies to remain longer and hunt more effectively in simple than in diverse agroecosystems, at least in those that are not too extensive.

Using a mathematical model analyzing the role of movement in the response of natural enemies to diversification, Corbett and Plant (1993) predicted that interplanted strip vegetation can act as either a sink or a source of natural enemies, depending on the mobility of the natural enemy and the specifics of system design. For a highly mobile predator, interplanted vegetation might be converted from a severe sink into a valuable resource simply by having interplanted vegetation germinate before the crop germinates, allowing early colonization by the predator. When the crop and interplanted vegetation germinate simultaneously, the model predicts that interplants act as a sink, but the magnitude of this effect will vary with natural enemy mobility, which determines spatial distributional patterns of predators and parasites.

Another shortcoming of the theory is that, so far, enemy and resource concentration hypotheses provide the basis to predict herbivore response in polycultures if

1. the herbivore is a specialist or exploits a narrow range of plants;
2. the polyculture is composed of a preferred host plant and one or more nonhost plants; or
3. host and nonhost plants overlap in time and space within the mixture.

Given these limitations, not all results from studies of agricultural diversification leading to reduced pest populations can be adequately explained by the two hypotheses as they stand. One such case is a study by Gold (1987) on the effects of intercropping cassava with cowpeas on the population dynamics of the cassava whiteflies (*Aleurotrachelus socialis* and *Trialeurodes variabilis*) in Colombia. As expected, intercropping reduced egg populations of both species of whitefly on cassava (a twelve-month crop), but the reductions were residual, persisting up to six months after the harvest of the cowpea, which lasts in the field only three to four months (Figure 3.4). The natural enemy hypothesis was rejected as a mechanism explaining reduced herbivore load, because predators were more abundant in monocultures than in polycultures, showing a numerical response. The resource concentration hypothesis could explain lower whitefly densities during the period that cowpea was present but cannot explain the reductions in whitefly populations observed long after the removal of the cowpea. Instead, it appeared that intercrop competition caused a reduction in cassava size or vigor that persisted through the remainder of the season. Thus, whitefly populations in poly- cultures were a function of host-plant selection and/or tenure time which, in turn, was related to host condition.

Another case is a study by Altieri and Schmidt (1986b), which found lower densities of the specialist herbivore *Phyllotreta cruciferae* on broccoli mixed with another crucifer host plant, the wild mustard *Brassica kaber.* Densities of *P. cruciferae* were greater on

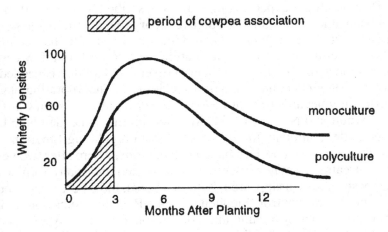

FIGURE 3.4. Expected population response (egg densities) of the whiteflies *Aleurotrachelus socialis* and *Trialeurodes variabilis* in cassava and cowpea mixtures in Colombia (after Altieri, 1991a).

broccoli plants grown in monocultures than in polycultures but not so on a per-plot basis. The differences in beetle abundance on a per-plant basis were basically due to the fact that the beetles concentrated more on wild mustard than on broccoli in the mixture.

This preference has a chemical basis since wild mustard has higher concentrations of glucosinolates than does broccoli, a strong beetle attractant. Thus, in this case, the differences in beetle abundance caused by diversity resulted from differences in beetle-feeding preferences rather than from differences in colonization, reproduction, or predation. Obviously, these trends do not conform to the assumptions of either of the two hypotheses.

These two case studies do not necessarily contradict the two hypotheses but instead provide additional explanations to current theory and call for more caution and flexibility since the responses of herbivores to vegetational diversity are not uniform and cannot always be explained by diversity per se. In fact, differences in tenure time or movement patterns between monocultures and polycultures are not always evident, although interplot differences in abundance persist. As suggested by the studies mentioned, other factors such as visual cues, microclimatic changes, feeding preferences, or direct effects on host-plant vigor could influence habitat location and/or searching behavior of both herbivores and natural enemies.

Despite all the experimental studies described here, we still have not been able to develop a predictive theory that enables us to determine what specific elements of biodiversity should be retained, added, or eliminated to enhance natural pest control. In a few cases, the simple addition of one element of diversity is all that is needed to ensure biological control of a pest species (e.g., incorporating blackberries in vineyards to control *Erythroneura* leafhoppers in California) (Doutt and Nakata, 1973). At times, all that is needed is to halt insecticide treatments to restore ecosystem regulating functions, as was the case in Costa Rican banana plantations, where after two years of nontreatment, major insect pests decreased and many former insect pests nearly disappeared. After ten years, most pests were under total biological control (Stephens, 1984). Recent studies comparing arthropod fauna in organic and conventional farming systems confirm the benefits of pesticide removal on the diversity of beneficial arthropods on the foliage and soils (Paoletti, Stinner, and Lorenzoni, 1989).

We also know little to suggest whether the ecological mechanisms proposed by the hypotheses work at a more regional level (i.e., at the level of agroecosystem mosaics interspersed among natural vegetation) and that pest problems will diminish due to spatial and temporal heterogeneity of agricultural landscapes. Some studies suggest that the vegetational settings associated with particular crop fields influence the kind, abundance, and time of arrival of herbivores and their natural enemies (Gut et al., 1982; Altieri and Schmidt, 1986a).

In perennial orchards (e.g., pear and apple) at temperate latitudes, a diverse complex of predators usually develops early in the season in orchards surrounded by woodlands. In these systems, main pests (e.g., *Psylla pyricola, Cydia pomonella,* etc.) are quickly reduced and maintained at low levels throughout the season. In contrast, early season predators are absent in more extensive commercial orchards, and therefore, pest pressure is more intense (Croft and Hoyt, 1983). Assuming that these trends also occur in the tropics, one would expect that in certain tropical agroecosystems (e.g., shifting cultivation in the lowland tropics), forest and bush fallows have potential value in controlling pests. Clearing small plots in a matrix of secondary forest vegetation may permit easy migration of natural enemies from the surrounding jungle (Matteson, Altieri, and Gagne, 1984). This positive potential role of natural vegetation on biological control is expected to change in view of current deforestation rates and modernization trends toward commercial monocultures.

Chapter 4

Insect Manipulation Through Weed Management

The presence of weeds within or around crop fields influences the dynamics of the crop and associated biotic communities. Studies over the past thirty years have produced a great deal of evidence that the manipulation of a specific weed species, a particular weed-control practice, or a cropping system can affect the ecology of insect pests and associated natural enemies (Van Emden, 1965b; Altieri, Schoonhoven, and Doll, 1977; Altieri and Whitcomb, 1979a,b; Thresh, 1981; William, 1981; Norris, 1982; Andow, 1983a). Weeds exert a direct biotic stress on crops by competing for sunlight, moisture, and some nutrients, thus reducing crop yields. Weeds indirectly affect crop plants through positive and/or negative effects on insect herbivores and also on the natural enemies of herbivores (Price et al., 1980). Herbivore-natural enemy interactions occurring in a crop system can be influenced by the presence of associated weeds or by the presence of herbivores on associated weed plants (Altieri and Letourneau, 1982). On the other hand, herbivores can mediate the interaction between crops and weeds, as in a natural community where the competitiveness of two plant species was altered substantially by the selective feeding of a foliage-consuming beetle (Bentley and Whittaker, 1979). Such relationships have been little explored in agricultural systems.

In this chapter, the multiple interactions among crops, weeds, herbivores, and natural enemies and, in particular, weed ecology and management that affect the dynamics of insect populations and thus crop health are discussed. The stress imposed on crops by weeds is viewed beyond the mere competitive interaction, incorporating an analysis of three trophic-level system interactions.

WEEDS AS SOURCES OF INSECT PESTS
IN AGROECOSYSTEMS

Weeds are important hosts of insect pests and pathogens in agro-ecosystems. Van Emden (1965b) cites 442 references relating to weeds as reservoirs of pests. One hundred such references concern cereals. A series of publications concerning weeds as reservoirs for organisms affecting crops has been published by the Ohio Agricultural Research and Development Center. More than seventy families of arthropods affecting crops were reported as being primarily weed associated (Bendixen and Horn, 1981). Many pest outbreaks can be traced to locally abundant weeds belonging to the same family as the affected crop plants. Many pest insects are sufficiently polyphagous that weeds unrelated to the crop may also be pest reservoirs. For example, *Aphis gossypii* feeds on over twenty unrelated weed species in and around cotton fields. Detailed examples of the role of weeds in the epidemiology of insect pests and plant diseases can be found in Thresh (1981), especially for crop diseases transmitted from weeds to adjacent crop plants by insect vectors. A major example provided is the role of the leafhopper *Circulifer tenellus* in transmitting the beet curly top virus from the western states of Utah and Colorado, where the vector itself is unable to survive the winter.

Weedy plants near crop fields can provide the requisites for pest outbreaks. The presence of *Urtica dioica* in the host layer of noncrop habitats surrounding carrot fields was the most important factor determining high levels of carrot fly larval damage to adjacent carrots (Wainhouse and Coaker, 1981). Adult leafhoppers invade peach orchards from edge vegetation and subsequently colonize trees under which groundcover is composed of preferred wild hosts (McClure, 1982). Plantains (*Plantago* spp.) provide alternative food for the rosy apple aphid *Dysaphis plantaginea,* an important pest of apple orchards in England. The rosy apple aphid spends most of the summer on plantains, returning to apples in late summer. The dock sawfly *Ametrastegia glabrata* normally feeds on docks (*Rumex* spp.) and knotgrass (*Polygonum* spp.), and the larvae of the last generation can move on to adjacent apple trees and bore into fruits or shoot tips (Altieri and Letourneau, 1982).

Certain grasses can act as hosts for cereal pests, and these species should be excluded from undersowings. *Bromus* spp., *Festuca* spp., and *Lolium multiflorum (italicum)* are among the hosts for *Sitobium avenae* and *Rhopalosiphum padi* which transmit the bean yellow dwarf virus (BYDV). Such species left as undersowings after harvest could act as a "green bridge" and encourage the transfer of the pest to the following crop. Less susceptible species are *Agropyron repens* and *Arrhenatherum elatius;* however, *A. repens* favors the buildup of saddle gall midge on *Lolium multiflorum* and is also highly susceptible to fruit fly, as are various *Festuca* and *Poa* spp., the latter also acting as hosts for wheat bulk fly (Burn, 1987).

Puvuk and Stinner (1992) report that the presence of grass weeds in cornfields is a factor that increases the attractiveness of those fields to the second flight of the European corn borer *(Ostrinia nubilalis).* Grassy vegetation appears to be preferred mating habitats for *O. nubilalis.*

Often the domesticated version of a plant species is the result of an intensive breeding program that results in, among other factors, a reduction in the concentration of secondary substances in various plant parts, producing plants with simpler and less stable defenses against herbivores (Harlan, 1975). The presence of wild relatives in crop borders may have local effects on the population genetics of a crop pest. Presumably, the wild plants can contribute to the genetic diversity of pest populations, at least if migration is limited. Thus, maintenance of a variety of plants that possess different complements of plant defenses in crop-border flora could result in the preservation of genetic diversity in the local pest population and decrease the rate of selection for new biotypes that have the ability to overcome host-plant resistance or withstand the application of pesticides (Thresh, 1981).

THE ROLE OF WEEDS IN THE ECOLOGY OF NATURAL ENEMIES

Certain weeds are important components of agroecosystems because they positively affect the biology and dynamics of beneficial insects. Weeds offer many important requisites for natural enemies such as alternative prey/hosts, pollen, or nectar, as well as microhabitats that are not available in weed-free monocultures (Van Emden, 1965b). Many insect pests are not continuously present in annual crops, and their predators and parasitoids must survive elsewhere during their absence. Weeds usually provide such resources (alternate host or pollen/nectar), aiding in the survival of viable natural enemy populations. The beneficial entomofauna associated with weeds has been surveyed for many species, including the perennial stinging nettle *(Urtica dioica)*, Mexican tea *(Chenopodium ambrosioides)*, camphorweed *(Heterotheca subaxillaris)*, and a number of ragweed species (Altieri and Whitcomb, 1979a). Perhaps the most exhaustive study of the fauna associated with various weeds is the work of Nentwig and associates in Berne, Switzerland, where they monitored the insects associated with eighty plant species sown as monocultures in a total of 360 plots (Nentwig, 1998). According to this survey, weed species are insect habitats of widely differing quality. Plants such as chervil of France *(Anthriscus cerefolium)*, comfrey *(Symphytum officinale)*, and gallant soldier *(Galinsoga ciliata)* have extremely low arthropod populations of less than 15 individuals/m^2, whereas most plants have 100 to 300 arthropods/m^2 according to the D-vac sampling method used by these researchers. Extremely high levels were found on poppy *(Papaver rhoeas)*, rape *(Brassica napus)*, buckwheat *(Fagopyrum esculentum)*, and tansy *(Tanacetum vulgare)*, where up to 500 or more arthropods were found per square meter. Considering the trophic structure of the arthropod communities, results were even more striking. Of all arthropods, phytophagous insects constituted about 65 percent of the species (most values between 45 and 80 percent), but the composition of the remaining arthropods varies greatly among predators and parasitoids and phytophagous arthropods or between aphidophagous predators and aphids according to the plant species. Most parasitoids were Hymenoptera of the families Aphidiidae, Braconidae, and Ichneumonidae, and also Proctotrupidae and Chalcidoidea, reaching about five to thirty individuals per square meter of vegetation, especially on Asteraceae and Brassicaceae weeds. Dominant predators included Empididae flies, Coleoptera (Cocinellidae, Carabidae, Staphilinidae, and Cantharidae), and chrysopid green lacewings. High densities (seventy predators/m^2) were observed on borage, blue knapweed, and *Papaver rhoeas*. Aphidophagous syrphids require a succession of plants, early- to late-flowering species, including highly attractive *Brassica, Sinapis,* and *Raphanus* species. Preferred oviposition sites for the common lacewing *(Chrysoperla carnea)* included about sixteen plants such as

Agrostemma githago, Trifolium arvense, Echium vulgare, Oenothera biennis, Centaurea jacea, and others (Zandstra and Motooka, 1978).

The Importance of Flowering Weeds

Most adult hymenopteran parasitoids require food in the form of pollen and nectar to ensure effective reproduction and longevity. Van Emden (1965b) demonstrated that certain Ichneumonidae, such as *Mesochorus* spp., must feed on nectar for egg maturation, and Leius (1967) reported that carbohydrates from the nectar of certain Unbelliferae are essential in normal fecundity and longevity in three Ichneumonid species. In studies of the parasitoids of the European pine shoot moth, *Rhyacionia buoliana,* Syme (1975) showed that fecundity and longevity of the wasps *Exeristes comstockii* and *Hyssopus thymus* were significantly increased with the presence of several flowering weeds. In Hawaii, *Euphorbia hirta* was reported as an important nectar source for *Lixophaga sphenophori,* a parasite of the sugarcane weevil (Topham and Beardsley, 1975). In San Joaquin Valley, California, *Apanteles medicaginis* wasps, parasites of the alfalfa caterpillar *(Colias eurytheme),* were often observed feeding on several weeds species *(Convolvulus, Helianthus,* and *Polygonum*), living longer and exhibiting a higher fecundity. Similar dependence on flowers have been reported for *Orgilus obscurator,* a parasite of the European pine shoot moth, and *Larra americana,* a parasite of the mole cricket (Zandstra and Motooka, 1978).

Wildflowers such as *Brassica kaber, Barbarea vulgaris,* and *Daucus carota* provided nectar flowers to female parasitoids of *Diadegma insulare,* an ichneumonid parasitoid of the diamondback moth (Idris and Grafius, 1995). An increased fecundity and longevity of the wasp was correlated with flower corolla opening diameter and flower shading provided to the parasitoid by the plants. Because of its long flowering period over the summer *Phacelia tanacetifolia* has been used as a pollen source to enhance syrphid fly populations in cereal fields in the United Kingdom (Wratten and Van Emden, 1995).

Spectacular parasitism increase has been observed in annual crops and orchards with rich undergrowths of wild flowers. In apple orchards, parasitism of tent caterpillar eggs and larvae and codling moth larvae was eighteen times greater in those orchards with floral undergrowths than in orchards with sparse floral undergrowth (Leius, 1967).

Soviet researchers at the Tashkent Laboratory (Telenga, 1958) cited lack of adult food supply as a reason for the inability of *Aphytis proclia* to control its host, the San Jose scale *(Quadraspidiotus perniciosus)*. The effectiveness of the parasitoid improved as a result of planting a *Phacelia* sp. cover crop in the orchards. Three successive plantings of *Phacelia* increased scale parasitization from 5 percent in clean-cultivated orchards to 75 percent where these nectar-producing plants were grown. These Soviet researchers also noted that *Apanteles glomeratus,* a parasite of two cabbageworm species *(Pieris* spp.) on crucifer crops, obtained nectar from wild mustard flowers. The parasites lived longer and laid more eggs when these weeds were present. When quick-flowering mustards were actually planted in the fields with cole crops, parasitization of the host increased from 10 to 60 percent (Telenga, 1958).

Weed flowers are also important food sources for various insect predators (Van Emden, 1965b). Pollen appears to be instrumental in egg production of many syrphid flies and is reported to be a significant food source for many predaceous Coccinellidae. Lacewings seem to prefer several composite flowers that supply nectar, thus satisfying their sugar requirements (Hagen, 1986).

Weeds may increase populations of nonpestiferous herbivorous insects (neutral insects) in crop fields. Such insects serve as alternative hosts or prey to entomophagous insects, thus improving the survival and reproduction of these beneficial insects in the agroecosystem. For example, the effectiveness of the tachinid *Lydella grisescens,* a parasite of the European corn borer, *Ostrinia nubilalis,* can be increased in the presence of an alternate host *Papaipema nebris,* a stalk borer on giant ragweed *(Ambrosia* spp.) (Syme, 1975). Several other authors have reported that the presence of alternate hosts on ragweeds near crop fields increased parasitism of specific crop pests. Examples include *Eurytoma tylodermatis* against the boll weevil, *Anthonomus grandis,* and *Macrocentrus ancylivorus* against the oriental fruit moth, *Grapholita molesta,* in peach orchards. The parasite *Herogenes* spp. uses the caterpillar of *Swammerdamia lutarea* on the weed *Crataegus* sp. to overwinter each year after emergence from the diamondback moth, *Plutella maculipennis* (Van Emden, 1965a). A similar situation occurs with the egg parasitoid *Anagrus epos,* whose effectiveness in regulating the grape leafhopper, *Erythroneura elegantula,* is greatly increased in vineyards near areas invaded by wild blackberry (*Rubus* sp.). This plant hosts an alternate leafhopper *Dikrella cruentata* that breeds in its leaves during winter (Doutt and Nakata, 1973). Stinging nettle *(Urtica dioica)* is a host of the aphid *Microlophium carnosum*. A large complex of predators and parasites attacks the aphid, the numbers of which increase rapidly in April and May in England before pest aphids appear on the crop plants. These natural enemies build up in numbers on this plant and then move on to adjacent crops once the nettles are cut in mid-June (Perrin, 1975).

Some entomophagous insects are attracted to particular weeds, even in the absence of host or prey, by chemicals released by the herbivore's host plant or other associated plants (Altieri et al., 1981). For example, the parasitic fly *Eucelatoria* sp. prefers okra to cotton, and the wasp *Peristenus pseudopallipes*, which attacks the tarnished plant bug, prefers *Erigeron* to other weed species (Monteith, 1960; Nettles, 1979). Parasitism by *Diaeretiella rapae* was much higher when the aphid *Myzus persicae* was on collard than when it was on beet, a plant lacking attractive mustard oil (Read, Feeny, and Root, 1970).

Of significant practical interest are the findings of Altieri and colleagues (1981), which showed that parasitization rates of *Heliothis zea* eggs by *Trichogramma* sp. were greater when the eggs were placed on soybeans next to corn and the weeds *Desmodium* sp., *Cassia* sp., and *Croton* sp. than on soybeans grown alone (Figure 4.1). Although the same number of eggs were placed on soybean and on the associated plants, few of the eggs placed on the weeds were parasitized, suggesting that these plants were not actively searched by *Trichogramma* sp. but nevertheless enhanced the efficiency of parasitization on the associated soybean plants. It is possible that they emitted volatiles with kairomonal action. Further tests showed that application of water extracts of some of these associated plants (especially *Amaranthus* sp.) to soybean and other crops enhanced parasitization of *H. zea* eggs by *Trichogramma* spp. wasps (Table 4.1). The authors stated that a stronger attraction and retention of wasps in the extract-treated plots may be responsible for the higher parasitization levels. The possibility that vegetationally complex plots are more chemically diverse than monocultures, and therefore more acceptable and attractive to parasitic wasps, opens new dimensions for biological control through weed management and behavior modification.

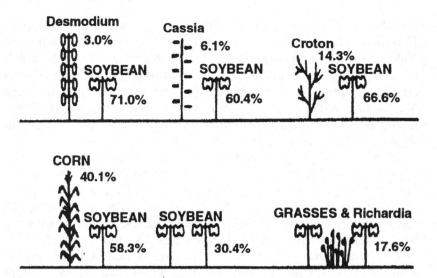

FIGURE 4.1. The effect of plant assemblages on parasitization of corn earworm *(Heliothis zea)* eggs by *Trichogramma* parasitic wasps in soybean fields (after Altieri and Letourneau, 1984).

In general, most beneficial insects present on weeds tend to disperse to crops, but in a few instances the prey found on weeds prevent or delay this dispersal. In such cases, allowing weeds to grow to ensure concentrations of insects and then cutting them regularly to force movement could be an effective strategy. For example, by cutting patches of stinging nettle *(U. dioica)* in May or June, predators (mainly Coccinellidae) were forced to move into crop fields (Perrin, 1975). Similarly, cutting the grass weed cover drove Coccinellidae into orchard trees in southeastern Czechoslovakia (Hodek, 1973). By cutting hedges of *Ambrosia trifida* infested with the weevil *Lixus scrobicollis,* a 10 percent increase of boll weevil parasitization by *E. tylodermatis* was obtained in two test plots of cotton adjacent to the hedgerow (Pierce, Cushman, and Hood, 1912). These practices must be carefully timed and based on the biology of beneficial insects. For example, in California the annual cleanup of weeds along the edges of alfalfa fields should be delayed until after mid-March, when aggregations of dormant Coccinellidae have largely dispersed (van den Bosch and Telford, 1964).

TABLE 4.1. Percent Parasitization of *Heliothis zea* (Boddie) Eggs by Naturally Occurring *Trichogramma* sp. in Crop Systems Sprayed with Various Plant Extracts

Treatment	Soybean	Cowpeas	Tomato	Cotton
Water extract of *Amaranthus*	21.4a[a]	45.4a	24.3a	13.6a
Water extract of corn	17.4b	45.8a	21.1a	—
Water	12.6c	31.6b	17.6b	4.2b

Source: After Altieri et al., 1981.

[a]Means followed by the same letter in each column are not significantly different according to Duncan's multiple range test ($P < 0.05$).

INSECT DYNAMICS IN WEED-DIVERSIFIED CROP SYSTEMS

In the past thirty-five years, research has shown that outbreaks of certain types of crop pests are less likely to occur in weed-diversified crop systems than in weed-free fields, mainly due to increased mortality imposed by natural enemies (Pimentel, 1961; Adams and Drew, 1965; Dempster, 1969; Flaherty, 1969; Smith, 1969; Root, 1973; Altieri, Schoonhoven, and Doll, 1977). Crop fields with a dense weed cover and high diversity usually have more predaceous arthropods than do weed-free fields (Pimentel, 1961; Dempster, 1969; Flaherty, 1969; Smith, 1969; Root, 1973; Perrin, 1975; Speight and Lawton, 1976). The successful establishment of several parasitoids has depended on the presence of weeds that provided nectar for the adult female wasps. Relevant examples of cropping systems in which the presence of specific weeds has enhanced the biological control of particular pests are given in Table 4.2 (Altieri and Letourneau, 1982). A literature survey by Baliddawa (1985) showed that population densities of twenty-seven insect species were reduced in weedy crops compared to weed-free crops. In addition to the insect studies, population densities of one mite species, *Eotetranychus willamette,* were found to be relatively lower in grape vines where weeds were allowed than in weeded plots (Flaherty, 1969). Table 4.3 lists the insect species under their respective orders and scores the number of occurrences that they appeared in the literature. The cabbage aphid and the cabbage white butterfly, *Pieris rapae,* seem to have received most attention.

TABLE 4.2. Selected Examples of Cropping Systems in Which the Presence of Weeds Enhanced the Biological Control of Specific Crop Pests

Cropping Systems	Weed Species	Pest(s) Regulated	Factor(s) Involved
Alfalfa	Natural blooming weed complex	Alfalfa caterpillar *(Colias eurytheme)*	Increased activity of the parasitic wasp *Apanteles medicaginis*
Alfalfa	Grass weeds	*Empoasca fabae*	Unknown
Apple	*Phacelia* sp. and *Eryngium* sp.	San Jose scale *(Quadraspidiotus perniciosus)* and aphids	Increased activity and abundance of parasitic wasps (*Aphelinus mali* and *Aphytis proclia*)
Apple	Natural weed complex	Tent caterpillar *(Malacosoma americanum)* and codling moth *(Cydia pomonella)*	Increased activity and abundance of parasitic wasps
Beans	Goosegrass *(Eleusine indica)* and red sprangletop *(Leptochloa filiformis)*	Leafhoppers *(Empoasca kraemeri)*	Chemical repellency or masking
Broccoli	Wild mustard	*Phyllotreta cruciferae*	Trap cropping
Brussels sprouts	Natural weed complex	Imported cabbage butterfly *(Pieris rapae)* and aphids *(Brevicoryne brassicae)*	Alteration of colonization background and increase of predators
Brussels sprouts	*Spergula arvensis*	*Delia brassicae*	Unknown
Brussels sprouts	*Spergula arvensis*	*Mamestra brassicae, Evergestis forficalis, Brevicoryne brassicae*	Increase of predators and interference with colonization
Cabbage	*Crataegus* sp.	Diamond moth *(Plutella maculipennis)*	Provision of alternate hosts for parasitic wasps (*Horogenes* sp.)
Citrus	*Hedera helix*	*Lachnosterna* spp.	Enhancement of *Aphytis lingnanensis*

Cropping Systems	Weed Species	Pest(s) Regulated	Factor(s) Involved
Citrus	Natural weed complex	Mites (*Eotetranychus* sp., *Panonychus citri, Metatetranychus citri*)	Unknown
Citrus	Natural weed complex	Diaspidid scales	Unknown
Coffee	Natural weed complex	Pentatomid *Antestiopsis intricata*	Unknown
Collards	Ragweed (*Ambrosia artemisiifolia*)	Flea beetle (*Phyllotreta cruciferae*)	Chemical repellency or masking
Collards	*Amaranthus retroflexus, Chenopodium album, Xanthium strumarium*	Green peach aphid (*Myzus persicae*)	Increased abundance of predators (*Chrysoperla carnea,* Coccinellidae Syrphidae)
Corn	Giant ragweed	European corn borer (*Ostrinia nubilalis*)	Provision of alternate hosts for the tachinid parasite *Lydella grisescens*
Corn	Natural weed complex	*Heliothis zea, Spodoptera frugiperda*	Enhancement of predators
Corn	*Setaria viridis* and *S. faberi*	*Diabrotica virgifera* and *D. barberi*	Unknown
Cotton	Ragweed	Boll weevil (*Anthonomus grandis*)	Provision of alternate hosts for the parasite *Eurytoma tylodermatis*
Cotton	Ragweed and *Rumex crispus*	*Heliothis* spp.	Increased populations of predators
Cotton	*Salvia coccinea*	*Lygus vosseleri*	Unknown
Cruciferous crops	Quick-flowering mustards	Cabbageworms (*Pieris* spp.)	Increased activity of parasitic wasps (*Apanteles glomeratus*)
Mungbeans	Natural weed complex	Beanfly (*Ophiomyia phaseoli*)	Alteration of colonization background

TABLE 4.2 *(continued)*

Cropping Systems	Weed Species	Pest(s) Regulated	Factor(s) Involved
Oil palm	*Pueraria* sp., *Flemingia* sp., ferns, grasses, and creepers	Scarab beetles *Oryctes rhinoceros* and *Chalcosoma atlas*	Unknown
Peach	Ragweed	Oriental fruit moth	Provision of alternate hosts for the parasite *Macrocentrus ancyclivorus*
Peach	Rosaceous weeds and *Dactylis glomerata*	Leafhoppers *Paraphelepsius* sp. and *Scaphytopius actus*	Unknown
Sorghum	*Helianthus* spp.	*Schizaphis graminum*	Enhancement of *Aphelinus* spp. parasitoids
Soybean	Broadleaf weeds and grasses	*Epilachana varivestis*	Enhancement of predators
Soybean	*Cassia obtusifolia*	*Nezara viridula, Anticarsia gemmatalis*	Increased abundance of predators
Soybean	*Crotalaria usaramoensis*	*Nezara viridula*	Enhancement of tachnid *Trichopoda* sp.
Sugar cane	*Euphorbia* spp. weeds	Sugar-cane weevil *(Rhabdoscelus obscurus)*	Provision of nectar and pollen for the parasite *Lixophaga sphenophori*
Sugar cane	Grassy weeds	Aphid *(Rhopalosiphum maidis)*	Destruction of alternate host plants
Sugar cane	*Borreria verticillata* and *Hyptis atrorubens*	Cricket *(Scapteriscus vicinus)*	Provision of nectar for the parasite *Larra americana*
Sweet potatoes	Morning glory *(Ipomoea asarifolia)*	Argus tortoise beetle *(Chelymorpha cassidea)*	Provision of alternate hosts for the parasite *Emersonella* sp.

Cropping Systems	Weed Species	Pest(s) Regulated	Factor(s) Involved
Vegetable crops	Wild carrot *(Daucus carota)*	Japanese beetle *(Popillia japonica)*	Increased activity of the parasitic wasp *Tiphia popilliavora*
Vineyards	Wild blackberry *(Rubus* sp. *)*	Grape leafhopper *(Erythroneura elegantula)*	Increase of alternate hosts for the parasitic wasp *Anagrus epos*
Vineyards	Johnson grass *(Sorghum halepense)*	Pacific mite *(Eotetranychus willamette)*	Buildup of predaceous mites *(Metaseiulus occidentalis)*

Source: Based on Altieri and Letourneau, 1982; Andow, 1991a.

Table 4.4 lists the main pest population regulating factors in the weed-diversified crops. These factors include parasites and predators, camouflage and masking, and reduced colonization. Natural enemies alone account for more than half (56 percent) of the cases where the pest population was claimed to be controlled.

Researchers have found at least two underlying mechanisms explaining how careful diversification of the weedy component of agricultural systems often lowers pest populations significantly (Altieri, Schoonhoven, and Doll, 1977; Risch, Andow, and Altieri, 1983). In some cases, plant dispersion and diversity appears to influence herbivore density, primarily by altering herbivore movement or searching behavior (Risch, 1981; Bach, 1980b; Kareiva, 1983). In other cases, predators and parasites encounter a greater array of alternative resources and microhabitats in weedy crops, reach greater abundance and diversity levels, and impose greater mortality on pests (Root, 1973; Letourneau and Altieri, 1983).

Many studies indicate that insect-pest dynamics are affected by the lower concentration and/or greater dispersion of crops intermingled with weeds. For example, adult and nymph densities of *Empoasca kraemeri,* the main bean pest of the Latin American tropics, were reduced significantly as weed density increased in bean plots. Conversely, the chrysomelid *Diabrotica balteata* was more abundant in weedy bean habitats than in bean monocultures; bean yields were not affected because feeding on weeds diluted the injury to beans.

TABLE 4.3. Occurrence of Individual Pests in Weed-Crop Diversity Studies

Crop Pest	Occurrence
Order: Coleoptera	
Anthonomus grandis Boheman	2
Popillia japonica Newm.	1
Phyllotreta striolata (Fabricius)	2
P. cruciferae (Goeze)	3
Rhaboscelus obscurus (Boisduval)	1
Order: Lepidoptera	
Anticarsia gemmatalis Hubner	1
Cydia pomonella (Linnaeus)	1
Colias eurytheme Boisduval	1
Evergestis forficalis Linnaeus	1
Grapholitha molesta (Busck)	1
Heliothis sp.	1
Malacosoma americanum (Fabricius)	1
Mamestra brassicae (Linnaeus)	1
Order: Lepidoptera	
Ostrinia nubilalis (Hubner)	1
Pieris rapae (Linnaeus)	8
Order: Diptera	
Ophiomyia (Melanogromyza) phaseoli (Tryon)	2
Order: Homoptera	
Aleyrodes brassicae (Walker)	1
Brevicoryne brassicae (Linnaeus)	7
Erythroneura elegantula (Osborn)	1
Erioischia brassicae (Linnaeus)	2
Empoasca kraemeri (Ross & Moore)	1
Lygus hesperus (Knight)	1
L. elisus (Van D.)	1
Myzus persicae (Sulzar)	4
Quadraspidiotus pernicious (Comstock)	1
Order: Heteroptera	
Nezara viridula Scudder	1
Order: Orthoptera	
Scapteriscus vicinus Scudder	1
Order: Acarina	
Eotetranychus willamette Ewing	1
Total	50

Source: After Baliddawa, 1985.

TABLE 4.4. Examples of Crop Pest Population Management Through Weed-Crop Diversity

	Agroecosystem	Pest	Factor (Suggested or Proved)	Reference
1	Cotton and cowpea strip planted with weeds	Bollweevil, *Anthonomus grandis*	Greater parasitic wasp, *Eurytoma* sp. population	Pierce (1912) quoted in Marcovitch (1935)
2	Vegetables grown among wild carrot (*D. carota*)	Japanese beetle, *Popillia japonica*	Greater activity of the parasitic wasp, *Tiphia popilliavora*	King and Holloway quoted by Altieri and Letourneau (1982)
3	Peach and ragweed (*Ambrosia* sp.), smart weed (*Polygonum* sp.) and lambs quarter (*Chenopodium album*), golden rod (*Solidago* sp.)	Oriental fruit moth, *Grapholitha molesta*	Alternate hosts for the parasite, *Macrocentrus ancylivorus*	Bobb (1939)
4	Sugar-cane with *Borreria verticillata* and *Hyptis atrorubens*	Cricket, *Scapteriscus vicinus*	Nectar source for the parasite, *Larra americana*	Wolcott (1942) quoted by Altieri and Letourneau (1982)
5	Apple trees grown with *Phacelia* sp. and *Bryngium* sp.	San Jose scale, *Quadraspidiotus perniciosus* and various species of aphids	Greater abundance and activity of the parasites, *Aphytis proclia*	Telenga (1958) quoted by Altieri and Letourneau (1982)
6	Collards, *Brassica oleracea*, and other brassicas grown among weeds	*Pyllotreta striolata, Myzus persiscae, Brevicoryne brassicae, Pieris rapae*	Camouflage	Pimentel (1960)
7	Cotton grown with ragweed *Ambrosia* sp.	Bollworm, *Anthonomus grandis*	Alternate hosts for the parasite, *Eurytoma tylodermis*	van den Bosch and Telford (1964)
8	Alfalfa with natural blooming weed complex	Alfalfa caterpillar, *Colias eurytheme*	Increased activity of *Apenteles medicaginis*	van den Bosch and Telford (1964)
9	Apple plants with weeds	Tent caterpillar, *Malacasoma americanum* and *Carpocapsa pomonella*	Increased activity and abundance of parasitic wasps	Lewis (1965)
10	Strip cropping cotton and alfalfa	Plant bugs, *Lygus herperus* and *L. elisus*	Retention of natural enemy and synchronizing intense natural enemy activity with pest population	van den Bosch and Stern (1969)

TABLE 4.4 *(continued)*

Agroecosystem	Pest	Factor (Suggested or Proved)	Reference
11 Grape vines with weeds	Pacific mite, *Eotetranychus williamettei*		Flaherty (1969)
12 Brussels sprouts with weeds (hoed or cut back to 15 cm)	*Myzus persicae, Brevicoryne brassicae, Aleyrodes brassicae, Pieris rapae*	Camouflage	Smith (1969) and (1976a)
13 Cruciferous crops with quick flowering mustards	Cabbage worm, *Pieris* sp.	Increased activity of the parasite, *Apanteles glomeratus*	National Academy of Sciences (1969)
14 Beans with goose grass, *Eleusine indica* and red spragletop, *Leptochloa filiformis*	Leafhopper, *Empoasca kraemeri*	Chemical repellency or masking	Tahvanainen and Root (1972)
15 Collards and rag-weed, *Ambrosia artemislifolia*	Flea beetle, *Phyllotreta cruciferae*	Chemical repellency and masking	Tahvanainen and Root (1972)
16 Cotton and rag-weed plus *Rumex crispus*	*Heliothis* sp.	Greater predator population	Smith and Reynolds (1972)
17 Vineyards with wild blackberry, *Rubus* sp.	Grape leafhopper, *Erythroneura elegantula*	More alternate hosts for *Anagrus epos*	Doutt and Nakata (1973)
18 Collards in "diverse" row in an old field	*P. cruciferae, M. periscae, B. brassicae, P. rapae*	Resource concen-tration	Root (1973)
19 Cabbage with white and red clover	*Erioischia brassicae, B. brassicae, P. rapae*	Less colonization and greater preda-tor population of *Harpalus rufipes, Phalangium* sp.	Dempster and Coaker (1974)
20 1, 10, 100 plant monocultures, 10 plant plots in old mowed field	*P. cruciferae, P. striolata,* aphids, and *P. rapae*	Camouflage	Cromartie (1981)
21 White clover undersown in brussels sprouts	*E. brassicae, B. brassicae, P. rapae*	Camouflage and greater predation	Dempster and Coaker (1974)
22 Mungbeans grown among weeds	Bean fly, *O. phaseoli*	Camouflage	Altieri and Whitcomb (1979b)

Agroecosystem	Pest	Factor (Suggested or Proved)	Reference
23 Corn grown with giant ragweed	European cornborer, *O. nubilalis*	Alternate hosts for *Lydella grisesens*	Syme (1975)
24 Sugar-cane with *Euphorbia* sp.	Sugar-cane weevil, *R. obscurus*	Nectar and pollen sources for *Lixophaga sphenophori*	Topham and Beardsley (1975)
25 Brussels sprouts grown among natural weeds complex	*P. rapae, B. brassicae*	Camouflage and more predation	Smith (1976a)
26 Mungbean grown among natural weed complex	Bean fly, *O. phaseoli*	Less colonization	Litsinger and Moody (1976)
27 Beans growing among weeds or surrounded by weedy borders	*E. kraemeri, D. balteata*		Altieri et al. (1977)
28 Brussels sprouts grown with *Spergula arvensis* weeds	*M. brassicae, E. forficalis, B. brassicae*	Lower colonization and greater predator population	Theunissen and den Ouden (1980)
29 Collards grown among weeds, mainly *Amaranthus retroflexus, Chenopodium album, Xanthium stramonium*	Green peach aphid, *M. persicae*	Greater abundance of predators, *Chrysopa camea*	Horn (1981)
30 Soybean grown with *Cassia obtusifolia*	The green stinkbug *N. viridula*, and velvet bean caterpillar *Anticarsia gemmatalis*	Greater abundance of predators	Altieri and Todd (1981)

Source: After Baliddawa, 1985.

In other experiments, *E. kraemeri* populations were reduced significantly in weedy habitats, especially in bean plots with grass weeds (*Eleusine indica* and *Leptochloa filiformis*). *D. balteata* densities fell by 14 percent in these systems. When grass-weed borders one meter wide surrounded bean monocultures, populations of adults and nymphs of *E. kraemeri* fell drastically (Figure 4.2). Also, pure stands of *L. filiformis* reduced adult leafhopper populations significantly more than *E. indica* (Schoonhoven et al., 1981); this effect ceased when the weed was killed with paraquat. If bean plots were sprayed with a wa-

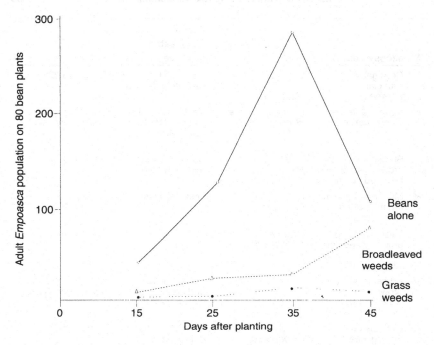

FIGURE 4.2. Effect of grass weed borders around 16 m² bean plots on the population of adult *Empoasca kraemeri* (after Altieri, Schoonhoven, and Doll, 1977).

ter homogenate of fresh grass-weed leaves, adult leafhoppers were repelled. Continuous applications affected the reproduction of leafhoppers, as evinced by a reduction in the number of nymphs (Altieri et al., 1977). Their regulatory effect was greater than that exhibited by extracts of broadleaf weeds such as *Amaranthus dubius.*

Weeds within a crop system can reduce pest incidence by enticing pest insects away from the crop. For example, flea beetles, *Phyllotreta cruciferae,* concentrate their feeding more on the intermingled *Brassica campestris* plants than on collards (Altieri and Gliessman, 1983). The weed species has significantly higher concentrations of allylisothiocyanate (a powerful attractant of flea beetle adults) than collards, thus diverting the beetles from the crops. Similarly, in Tlaxcala, Mexico, the presence of flowering *Lupinus* spp. in tasseling cornfields often diverts the attack of the scarab beetle, *Macrodactylus* sp., from female corn flowers to lupine flowers (Trujillo-Arriaga and Altieri, 1990).

Several studies have documented pest reduction due to an increase of natural enemies in weedy crop fields. Fall armyworm, *Spodoptera frugiperda,* incidence was consistently higher in weed-free corn plots than in corn plots containing natural weed complexes or selected weed associations. Corn earworm *(Heliothis zea)* damage was similar in all weed-free and weedy treatments, suggesting that this insect is not affected greatly by weed diversity. Experimental design was a crucial factor in demonstrating the effect of weeds on predator populations. In an experiment conducted by Altieri and Whitcomb (1980), field plots were close together (8 m apart), and predators moved freely between habitats, confounding results. Therefore, it was difficult to identify between-treatment differences (i.e., weedy versus weed-free plots) in the composition of predator communities. In another experiment, increased distances (50 m) between plots minimized such migrations, resulting in greater population densities and diversity of common foliage insect predators in the weed-manipulated corn systems than in the weed-free plots. Trophic relationships in the weedy habitats were more complex than food webs present in monocultures.

When comparing parasitism of second generation *O. nubilalis* larvae by the ichneumonid *Eriborus terebrans,* Puvuk and Stinner (1992) found that parasitism was not significantly influenced by the presence of weeds, although there was a trend for greater parasitism in treatments with weeds than in weedless plantings.

In England, winter barley plots with grass weeds had fewer aphids and more than ten times the number of staphylinid beetles than plots without weeds (Burn, 1987). Similarly, spring-planted alfalfa plots infested with weeds had a less diverse substrate-predator complex but a greater foliage-predator complex than did weed-free plots (Barney et al., 1984). The carabid *Harpalus pennsylvanicus* and foliage predators (i.e., *Orius insidiosus* and Nabidae) were more abundant in alfalfa fields where grass weeds were dominant.

Smith (1969) concluded that weeds within brussels sprouts enhanced natural enemy action against aphids by providing predator oviposition sites. This partly explained the lower aphid populations recorded in weedy plots. By selectively allowing a cover of *Spergula arvensis* within brussels sprouts plots, populations of *Mamestra brassicae, Evergestis forficalis,* cabbage root fly, and *Brevicoryne brassicae* were drastically reduced (Theunissen and den Ouden, 1980).

Schellhorn and Sork (1997) compared population densities of herbivores, predators, and parasitoids on collards in monocultures and on collards interplanted with two different groups of weeds, one with weed species from the same plant family as the collards (Brassicaceae) and one with weed species from unrelated plant families (non-Brassicaceae). The collards in the Brassicaceae-weed polyculture had higher densities (number of herbivores/mean leaf area [cm^2] per plant) of specialist herbivores than collards in the non-Brassicaceae-weed polyculture and in collard monoculture. The resource concentration hypothesis is supported by the observation of higher populations of *Phyllotreta* spp., acting as facultative polyphages, in the Brassicaceae-weed polyculture than in the non-Brassicaceae-weed polyculture where *Phyllotreta* spp. are facultative monophages. Population densities of natural enemies (mostly coccinelids, carabids, and staphylinids) were higher in the polycultures than in the monoculture: carabid and staphylinid predators may be responsible for the observed larval mortality in the imported cabbage worm, *Pieris rapae,* and in the diamondback larvae, *Plutella xylostella.* In spite of differences in densities of specialist herbivores across treatments, crop yield, leaf area (cm^2), the proportion of leaf area damaged, and the number of leaves undamaged did not differ. The authors concluded that the use of weedy cultures can provide effective means of reducing herbivores if the crop and weed species are not related and plant competition is prevented.

Based on his extensive studies of weeds as sources of habitat for natural enemies, Nentwig (1998) prepared seed mixtures consisting of twenty-four species of wildflowers sown as 3- to 8-m-wide weed strips planted every 50 to 100 m within fields. Those strips are referred to as ecological compensation areas, which serve as a refuge area and/or dispersal center of natural enemies, compensating, at least partially, for the negative effects of monoculture (Frank and Nentwing, 1995). These studies have demonstrated enhanced biodiversity of beneficial insects and lower insect-pest incidence in crop fields enriched with weed strips.

ISOLATING THE ECOLOGICAL EFFECTS
OF WEED DIVERSITY

Although all hypotheses explaining herbivore population dynamics in weed-diversified cropping systems attribute a role to the physical structure of the plant canopies in achieving decreased herbivore abundance and/or increased natural enemy densities, few experiments have removed the confounding influence of plant-species richness to assess the effects of plant architecture and density per se (Altieri and Letourneau, 1984). One problem with these experiments is that they do not isolate crop-weed diversity as an independent variable. In most cases, weed diversity could possibly have reduced herbivore abundance because it reduced the size or quality of crop plants (Kareiva, 1983). Weed density, diversity, plot, or patch sizes are all interacting factors that may influence crop quality and herbivore densities.

The presence of weeds in crops affects both plant density and spacing patterns, factors known to significantly influence insect populations (Mayse, 1983). In fact, many herbivores respond specifically to plant density; some proliferate in close plantings, whereas others reach high numbers in open-canopy crops. Predator and parasite populations tend to be greater in high-density plantings. Mayse suggests that the microclimate associated with canopy closure, which occurs earlier in dense plantings, may increase development rates of some predators and possibly facilitate prey capture.

Careful consideration of the units to be used in expressing population numbers is crucial for meaningful interpretations of results and for determining general patterns among various research findings. For example, Mayse and Price (1978) found that numbers of certain arthropod species sampled in different soybean row-spacing treatments were significantly different on a per-plant basis, but those same population values converted to a square meter of soil-area basis were not significantly different.

Based on the preceding considerations, in addressing the effects of crop-weed density/spacing on insect populations, one should fully consider (1) the effects on the growth, development, and nutritional status of the crop plants and weeds, (2) the effects on the microclimate and microhabitats available for the life processes of herbivores and their natural enemies, and (3) the effects of potentially different levels of herbivores in the population dynamics of predators and parasites.

Andow (1991b) purposely designed an experiment to separate the effects of decreased pest attack (from *Epilachna varivestis* and *Empoasca fabae*) and increased plant competition that often occur simultaneously in weed-diversified bean systems. In this system, weeds directly affected bean herbivores by reducing their population densities and intensity of attack, indirectly benefiting the bean plants. At the same time, the weeds directly competed with the beans. There is a negative correlation between the intensity of insect attack on beans and the intensity of competition. At one extreme, in the monoculture, there was very high pest attack but no interspecific competition, whereas at the other extreme, the unweeded treatment, there was very intense competition but very little insect attack. Without insects, competition reduced the yield of beans in these treatments. However, when insects were present, there was no difference in yield among the three treatments. Thus, at low levels of crop-weed competition, the effects of reduced insect-pest damage were large enough to balance out the effects of increased plant competition (Figure 4.3).

In contrast, there was no significant interaction between yields and insect damage in the unweeded treatment. This was the treatment with the greatest reduction in insect-pest populations and herbivore damage. Despite this great decrease in the intensity of herbivory, there was no detectable yield response. Apparently, the reduction in yield from the intense weed competition was so large that the positive effect of reduced herbivory was swamped out. Andow (1991b) concluded that three-way interactions among beans, weeds, and bean herbivores were important when bean-weed competition was not very intense.

CROP-WEED MANAGEMENT CONSIDERATIONS

As indicated by the studies discussed in this chapter, much evidence suggests that encouragement of specific weeds in crop fields may improve the regulation of certain insect pests (Altieri and Whitcomb, 1979a). Naturally, careful manipulation strategies need to be defined in order to avoid weed competition with crops and interference with certain cultural practices. Moreover, economic thresholds

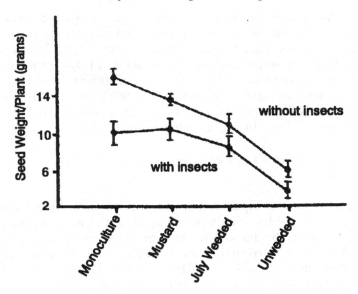

FIGURE 4.3. Yields of beans (as seed weight per plant) with and without herbi-vores and with and without weeds. Insects were eliminated with insecticides. The four levels of weediness were (1) no weeds (monoculture), (2) interplanted with 8,200 clumps of wild mustards *(Brassica kaber)*/ha midway between alter-nate bean rows, (3) natural populations of weeds for the first thirty-five days after planting and weeded from then on (July weeded), and (4) natural populations of weeds for the entire growing season (unweeded) (after Andow, 1991b).

of weed populations, as well as factors affecting crop-weed balance within a crop season, need to be defined (Bantilan, Palada, and Har-wood, 1974).

Shifting the crop-weed balance so that insect regulation is achieved and crop yields are not economically reduced may be accomplished by carefully using herbicides or selecting cultural practices that favor the crop cover over weeds. Suitable levels of desirable weeds that support populations of beneficial insects can be attained within fields by (1) designing competitive crop mixtures, (2) allowing weed growth as alternate rows or in field margins only, (3) using cover crops, (4) adopting close-row spacing, (5) providing weed-free periods (i.e., keeping crops free of weeds during the first third of their growth cy-cle), (6) mulching, (7) practicing soil-fertility management, and (8) practicing cultivation regimes.

In addition to minimizing the competitive interference of weeds, changes in the species composition of weed communities are desirable to ensure the presence of plants that attract beneficial insects. Manipulation of weed species can be achieved by several means, such as changing levels of key chemical constituents in the soil, use of herbicides that suppress certain weeds while encouraging others, direct sowing of weed seeds, and timing soil disturbances (Altieri and Whitcomb, 1979b; Altieri and Letourneau, 1982).

Changes of the Levels of Key Chemical Constituents in the Soil

The local weed complex can be affected indirectly by the manipulation of soil fertility. Fields in Alabama, with low soil potassium, were dominated by buckhorn plantain *(Plantago lanceolata)* and curly dock *(Rumex crispus),* whereas fields with low soil phosphorus were dominated by showy crotalaria *(Crotalaria spectabilis),* morning glory *(Ipomoea purpurea),* sicklepod *(Cassia obtusifolia), Geranium carolinianum,* and coffee senna *(Cassia occidentalis)* (Hoveland, Buchanan, and Harris, 1976). Soil pH can influence the growth of certain weeds. For example, weeds of the genus *Pteridium* occur on acid soils while *Cressa* sp. inhabits only alkaline soils. Other species (many Compositae and Polygonaceae) can grow in saline soils (National Academy of Sciences, 1969).

Other major soil-management practices that affect soil processes related to soil-weed dynamics are tillage, crop rotation, and use of cover crops and green manures. Combined in a cropping system these practices (1) can reduce the persistence of weed seeds in the soil, (2) reduce the abundance of safe-sites and the filling of available sites, and (3) reduce individual crop-yield loss per individual weed (Liebman and Gallandt, 1997).

Use of Herbicides

Repeated herbicide treatments can cause a shift in weed populations or select for the development of resistant weed biotypes at the expense of susceptible community members (Horowitz et al., 1962).

Buchanan (1977) has published a list of herbicides that suppress certain weeds while encouraging others. When a maximum rate of 0.6 kg/ha of trifluralin (a,aa-trifluoro-2, 6-dinitro-N,N-dipropylptoluidine) is applied before sowing, populations of velvetleaf *(Abutilon theophrasti)*, jimson weed *(Datura stramonium)*, venice mallow *(Hibiscus trionum)*, and prickly sida *(Sida spinosa)* can be grown among cotton and soybeans without the presence of other unwanted weed species. Although most examples cited by Buchanan (1977) relate to weed-control studies, similar methods may be developed to favor particular beneficial weeds in order to achieve early increases of natural enemy populations. Of course such an approach has no applicability in organic farming systems.

Direct Sowing

Direct sowing of the grasses *Eleusine indica* and *Leptochloa filiformis* to form a one-meter border around bean fields in Colombia, decreased colonization and reproductive efficiency in the leafhopper *Empoasca kraemeri* (Altieri and Whitcomb, 1979a). The application of this method, however, demands careful investigation of certain weed-seed germination requirements. Some seeds remain in enforced dormancy and germinate only under certain environmental conditions. Most weed seeds have specialized requirements for germination, making it difficult to sow weeds for experimental purposes (Anderson, 1968). Nevertheless, today in the market it is possible to find many weed-seed mixtures (mostly flowering plants) that are recommended for planting in and around crop fields to create habitats for beneficial insects.

Soil Disturbance

The weed-species composition of recently cropped fields can be manipulated by changing the season of disturbance. In northern Florida, field plots plowed at different times of year exhibited different weed-species composition. Within these plots, populations of herbivorous insects fluctuated according to composition and abundance of weed hosts. Large numbers of chrysomelids and leafhoppers were collected in treatment plots where preferred weed hosts reached high cover values. As these herbivores served as alternative prey, the number of predaceous arthropods feeding on them varied in direct proportion to the size of populations of their preferred herbivorous prey as determined by the presence of weed hosts and the season of plowing

(Altieri and Whitcomb, 1979a). The authors proposed plowing strips of land within a crop in different seasons to encourage specific weeds that, in turn, provide an alternative food and habitat to specific predators (Table 4.5). If this is done early in the season, a balance of natural enemies can be maintained in the field, before outbreaks of pest species occur.

Modifying Weed Spatial Patterns

It may be possible to influence weed spatial distributions and promote weeds to occur in clumps within fields rather than being uniformly distributed. For a given average density over a broad area, clumped weeds are expected to be less damaging to crop yield than are randomly or evenly distributed weeds (Liebman and Gallandt, 1997). Clumped weeds in a field spot may reduce yields there but provide a source of beneficials that colonize the rest of the fields from the clump.

Manipulating the Weed's Critical Competition Period

Perhaps one of the most useful concepts in managing weeds for insect regulation within fields is the "critical period." This is the maximum period that weeds can be tolerated without affecting final crop yields or the point after which weed growth does not affect final yield. In general, weeds that emerge earlier in the growing season are more damaging to crop yields than are populations that emerge later. On the other hand, crops differ in their sensitivity to different durations of weed competition, but most are most susceptible during the first third of their life cycle. The guiding principle here is to delay weed emergence relative to crop emergence (Liebman and Gallandt, 1997) Duration of weed-competition data for particular crops has been compiled by Zimdahl (1980) and critical weed-free maintenance periods have been identified for various crop-weed associations. The important question becomes how long exclusion efforts must be maintained before they can be relaxed, so that weeds emerge and provide the desired entomological benefits. As might be expected, the critical weed-free period for a given crop varies considerably among sites and years, due to climate and edaphic conditions affecting crop and weed emergence and growth rates, weed-species composition, and weed density.

TABLE 4.5. Selected Examples of Associations Between Herbivorous Insects and Predaceous Arthropods Occurring on Specific Weed Species Which Respond to Particular Dates of Soil Disturbance in North Florida

Weed Species	Date(s) of Soil Disturbance That Enhances the Weed Population	Herbivore(s) Associated with the Weed That Serve As Alternate Prey to Various Predators	Predaceous Arthropods That Feed on the Herbivore(s) Associated with the Weed
1 *Oenothera laciniata* (early evening primrose)	August	*Altica* sp. (leaf beetle)	*Lebia viridus* (ground beetle)
2 *O. biennis* (evening primrose)	December	*Altica* sp.	*L. viridus*
3 *Amaranthus* sp. (pigweed)	April	*Disonycha glabrata* (leaf beetle)	*Lebia analis* (ground beetle)
4 *Heterotheca subaxillaris* (camphorweed)	October	*Zygogramma heterothecae* (leaf beetle) and 30 other herbivorous insect species	30 predaceous arthropod species
5 *Chenopodium ambrosiodes* (Mexican tea)	December	*Z. suturalis* and 31 other herbivorous insect species	34 predaceous arthropod species
6 *Solidago altissima* (goldenrod)	December	*Uroleucon* spp. (11 aphid species) and 28 other herbivorous species	58 predaceous arthropod species
7 *Ambrosia artemisiifolia* (ragweed)	December October	*Z. suturalis, Nodonota* sp. (leaf beetles) *Reuteroscopus ornatus* (plant bug) *Uroleucon ambrosiae* (aphid) *Epiblema* sp., *Tarachidia* sp. (caterpillars) and 17 other herbivorous insect species	*Cycloneda sanguinea* (lady beetle) *Zelus cervicalis* and *Sinea* spp. (assassin bugs) *Peucetia viridans* (lynx spider) and 4 other predaceous arthropod species
8 *Lactuca canadensis* (wild lettuce)	Control (not disked)	*Uroleucon* sp. (aphids)	*Podabrus* sp. (soldier beetle) *Cycloneda sanguinea Chrysopa* sp. (lacewing) *Doru taeniatum* (earwig) *Syrphids* and spiders

Source: After Altieri and Letourneau, 1982.

In studies in southern Georgia, Altieri et al. (1981) observed that populations of the velvetbean caterpillar *(Anticarsia gemmatalis)* and of the southern green stink bug *(Nezara viridula)* were greater in weed-free soybeans than in soybeans left weedy for either two or four weeks after crop emergence or for the whole season. Soybeans maintained weed free for two or four weeks after emergence required no further weed control to produce optimum yield (Walker et al., 1984).

In another experiment conducted in California, allowing weed growth during selected periods of the collard crop cycle (two or four weeks weed free or weedy all season) resulted in lower flea beetle *(P. cruciferae)* densities in the weedy monocultures than in the weed-free monocultures. Lowest densities occurred in systems allowed to remain weedy all season. No differences in the abundance of beetles were observed between collards kept weed free for two or four weeks after transplanting (Altieri and Gliessman, 1983). Beetle densities were lower in these systems than in the weed-free system.

In collard systems with "relaxed" weeding regimes, flea beetle densities were at least five times greater on a per-plant basis on *Brassica campestris* (the dominant plant of the weed community) than on collards. *B. campestris* germinated quickly and flowered early, each plant averaging a height of 39 cm with twelve leaves and sixteen open flowers, sixty days after germination. This apparent preference of *Phyllotreta cruciferae* for *B. campestris* over collards resulted in a higher concentration of flea beetles on the wild crucifer, diverting flea beetles from collards and consequently diluting their feeding on the collards (Table 4.6). Collards grown under various levels of weed diversity exhibited significantly less leaf damage than monoculture collards grown in weed-free situations (Altieri and Gliessman, 1983). In a similar experiment, wild mustard sowed one week after broccoli transplanting showed no reduction of broccoli yield and reduced aphid numbers while increasing effective predation by syrphid larvae (Kloen and Altieri, 1990).

Defining periods of weed-free maintenance in crops so that numbers of pests do not surpass tolerable levels might prove to be a significant compromise between weed science and entomology, a necessary step to further explore the interactions described in this chapter.

Unquestionably, weeds stress crop plants through interference processes. However, substantial evidence suggests that weed presence in crop fields cannot be automatically judged damaging and in need of

TABLE 4.6. Mean Flea Beetle (*Phyllotreta cruciferae*) Densities, Weed and Crop Biomass in Various Collard Cropping Systems in Santa Cruz, California

Cropping System	No. of flea beetles		Leaves in Each Collard Plant with Beetle Damage (45 Days After Transplanting) (%)	Weed Biomass (g/m²)	Whole-Plant Collard Dry Weight (g/m²)
	Per 10 Collard Plants	Per 5 *Brassica campestris* Plants			
Weed free all season	34.0 ± 2.6a	—	54.4a	0	213.6
Weed free for 4 weeks after collard transplanting	29.3 ± 1.7b	78.1 ± 16.3a	44.6b	52.3	361.3
Weed free for 2 weeks after collard transplanting	29.0 ± 1.7b	73.7 ± 20.1a	44.5b	55.2	243.0
Weedy all season	6.6 ± 3.8c	25.0 ± 11.5b	29.9c	483.2	226.1

Source: After Altieri and Gliessman, 1983.

Note: Means followed by the same letter in each column are not significantly different ($P = 0.05$). All means are averages of the three sampling dates (15, 30, and 45 days after transplanting)

75

immediate control. In fact, crop-weed interactions are overwhelmingly site specific and vary according to plant species involved, environmental factors, and management practices. Thus, in many agroecosystems, weeds are ever-present components adding to the complexity of interacting trophic levels mediating a number of crop-insect interactions with major effects on final yields. It is here argued that in weed-diversified systems we cannot understand plant-herbivore interactions without understanding the effects of plant diversity on natural enemies, nor can we understand predator-prey and parasite-host interactions without understanding the role of the plants involved in the system.

An increasing awareness of these ecological relationships should place emphasis on weed management, as opposed to weed control, so that herbicides may be considered merely a component part of a total system for managing weeds, where season-long, weed-free monocultures are not always assumed to be the best crop-production strategy (Aldrich, 1984).

Chapter 5

Insect Management
in Multiple-Cropping Systems

The single-species nature of crop systems can be broken by growing crops in polycultural patterns. Polycultures are systems in which two or more crops are usually planted simultaneously and sufficiently close together to result in interspecific competition and/or complementation. These interactions may have inhibiting or stimulatory effects on yields (Hart, 1974). In the design and management of these systems, one strategy is to minimize competition and maximize complementation among species in the mixture (Francis, Flor, and Temple, 1976). Among the potential advantages that can emerge from the intelligent design of polycultures are population reduction of insect pests, suppression of weeds through shading by complex canopies or allelopathy (Gliessman and Amador, 1980), better use of soil nutrients (Igzoburkie, 1971), and improved productivity per unit of land (Harwood, 1974).

In the tropics, polycultures are an important component of tropical small-farm agriculture and, in addition to lowering risks, one of the reasons for the evolution and adoption of such cropping patterns may be the reduced incidence of insect pests (Altieri and Liebman, 1988; Alghail, 1993). Polyculture systems may also provide a potential for improved crop productivity even in temperate agriculture. In fact, the systems are so prevalent that quantitative estimates suggest that 98 percent of the cowpeas grown in Africa and 90 percent of the beans in Colombia are intercropped. The percentage of cropped land in the tropics actually devoted to intercropping varies from a low of 17 percent in India to a high of 94 percent in Malawi. Apparently, in El Salvador, it used to be impossible to find sorghum that was not intercropped with maize. Even in temperate North America, before the widespread use of modern varieties and mechanization, intercropping was apparently common (e.g., 57 percent of the soybean acreage in

77

Ohio was grown in combination with maize in 1923) (Vandermeer, 1989). In the Midwest of the United States, the combination of soybean and corn in strip intercropping has been tried as an economic alternative to monocultures (Francis, 1990).

Polyculture management basically consists of the design of spatial and temporal combinations of crops in an area. The arrangement of crops in space can be in the form of such systems as strip cropping, intercropping, mixed-row cropping, and cover cropping (Andrews and Kassam, 1976). The crop arrangement in time can vary according to whether mixed crops are planted simultaneously or in sequence as rotational crops, relay crops, or ratoon crops, or whether crops are combined in a synchronous or an asynchronous fashion or in a continuous or discontinuous planting pattern (Litsinger and Moody, 1976). According to Francis, Flor, and Temple (1976), desirable features of crops to be considered for intercropped systems include photoperiod insensitivity, early and uniform maturity, low stature and nonlodging effects, good·population response, insect and disease resistance, efficient soil-fertility response, and high yield potential.

Multiple-cropping systems constitute agricultural systems diversified in time and space. As mentioned earlier, much evidence suggests that this vegetational diversity often results in a significant reduction of insect-pest problems (Altieri and Letourneau, 1982). A large body of literature cites specific crop mixtures that affect particular insect pests (Litsinger and Moody, 1976; Perrin, 1977; Perrin and Phillips, 1978; Andow, 1983a), and other papers explore the ecological mechanisms involved in pest regulation (Root, 1973; Bach, 1980a,b; Risch, 1981). In polycultures, apart from the evident increase in plant-species diversity, there are changes in patch size, plant density, and plant quality. All of these changes affect density of pests and other organisms. Clearly, much knowledge has accumulated, and this acquired information is slowly providing a basis for designing complex crop systems so that pest problems and the need for active control measures are minimized (Murdoch, 1975).

PATTERNS OF INSECT ABUNDANCE IN POLYCULTURES

In recent years, agroecologists have conducted experiments in multiple-cropping systems to test the theory that increased plant diversity fosters stability of insect populations (Pimentel, 1961; Root, 1973; Van Emden and Williams, 1974). A recent examination of all available studies on the effects of these patterns on insect-pest populations tends to support the theory, although confusion may arise depending on how diversity and stability are defined (Risch, Andow, and Altieri, 1983). In multiple cropping, structural and species vegetational diversity (a measure of the biotic, structural, and microclimatic complexity arising from the mixing of different plants) results from the addition of crop species in time and space. Stability herein refers to low pest-population densities over time.

Examples of specific crop mixtures that result in reduced pest incidence can be found in Litsinger and Moody (1976), Altieri and Letourneau (1982), Andow (1983b), and Altieri and Liebman (1988) and are summarized in Table 5.1. Thirty-five insect species were investigated in fifty insect studies. The majority of the insects were in the orders Lepidoptera, Coleoptera, and Homoptera, accounting for 42, 32, and 18 percent, respectively, of the total crop pests. A combination of lowered resource concentration, trap cropping, various diversionary mechanisms, planting density, and plant physical obstruction account for 22.5 percent of the factors explaining pest reduction. Predators and parasites account for only 15 and 10 percent, respectively, whereas masking and/or camouflaging and repellency account for 12.5 percent each. Overall natural enemy action was responsible for up to 30 percent of the control of the studied crop pests, and the remaining known cases were controlled by other factors.

By analyzing a series of case studies, Helenius (1991) showed that monophagous insects are more susceptible to crop diversity than are polyphagous insects and cautioned the increased risk of pest attack if the dominant herbivore fauna in a given agroecosystem is polyphagous. The reduction in pest numbers for monophagous insects was almost twice (53.5 percent of the case studies showed lowered numbers in polycultures) that for polyphagous insects (33.3 percent).

Coll (1998) compared parasitoid density and parasitism rates in forty-two different monoculture-intercrop systems reported in the literature. In two-thirds of the comparisons, the parasitoids were more abundant or attacked more hosts in the intercropped than monocultured habitats. However, in about one-third of the comparisons, no consistent differences were recorded in parasitoid density or parasitism rate among habitats. A lower rate of parasitoid enhancement was

TABLE 5.1. Selected Examples of Multiple-Cropping Systems That Effectively Prevent Insect-Pest Outbreaks

Multiple-Cropping System	Pest(s) Regulated	Factor(s) Involved
Beans grown in relay intercropping with winter wheat	*Empoasca fabae* and *Aphis fabae*	Impairment of visual searching behavior of dispersing aphids
Brassica crops and beans	*Brevicoryne brassicae* and *Delia brassicae*	Higher predation and disruption of oviposition behavior
Brussels sprouts intercropped with fava beans and/or mustard	Flea beetle *Phyllotreta crucifecae* and cabbage aphid *Brevicoryne brassicae*	Reduced plant apparency trap cropping, enhanced biological control
Cabbage intercropped with white and red clover	*Erioischia brassicae*, cabbage aphids, and imported cabbage butterfly *(Pieris rapae)*	Interference with colonization and increase of ground beetles
Intercropping of *Cajanus cajan* with red, black, and green gram	Pod borers, jassids, and membracids	Delayed colonization of herbivores
Cassava intercropped with cowpeas	Whiteflies *Aleurotrachelus socialis* and *Trialeurodes variabilis*	Changes in plant vigor and increased abundance of natural enemies
Cauliflower strip cropped with rape and/or marigold	Blossom beetle *Meligethes aeneus*	Trap cropping
Corn intercropped with beans	Leafhoppers *(Empoasca kraemeri)* leaf beetle *(Diabrotica balteata)* and fall armyworm *(Spodoptera frugiperda)*	Increase in beneficial insects and interference with colonization
Corn intercropped with fava beans and squash	Aphids, *Tetranychus urticae* and *Macrodactylus* sp.	Enhanced abundance of predators
Corn intercropped with clover	*Ostrinia nubilalis*	Unknown
Corn intercropped with soybean	European corn borer *Ostrinia nubilalis*	Differences in corn varietal resistance
Corn intercropped with sweet potatoes	Leaf beetles *(Diabrotica* spp.) and leafhoppers *(Agallia lingula)*	Increase in parasitic wasps

Multiple-Cropping System	Pest(s) Regulated	Factor(s) Involved
Intercropping corn and beans	*Dalbulus maidis*	Interference with leafhopper movement
Cotton intercropped with forage cowpea	Boll weevil *(Anthonomus grandis)*	Population increase of parasitic wasps (*Eurytoma* sp.)
Intercropping cotton with sorghum or maize	Corn earworm *(Heliothis zea)*	Increased abundance of predators
Cotton intercropped with okra	*Podagrica* sp.	Trap cropping
Strip cropping of cotton and alfalfa	Plant bugs *(Lygus hesperus* and *L. elisus)*	Prevention of emigration and sychrony in the relationship between pests and natural enemies
Strip cropping of cotton and alfalfa on one side and maize and soybean on the other	Corn earworm *(Heliothis zea)* and cabbage looper *(Trichoplusia ni)*	Increased abundance of predators
Intercropping cowpea and sorghum	Leaf beetle *(Oetheca benningseni)*	Interference of air currents
Cucumbers intercropped with maize and broccoli	*Acalymma vittata*	Interference with movement and tenure time on host plants
Groundnuts intercropped with field beans	*Aphis craccivora*	Aphids trapped on epidermal hairs of beans
Maize intercropped with canavalia	*Prorachia daria* and fall armyworm *(Spodoptera frugiperda)*	Not reported
Maize-bean intercropping	*Spodoptera frugiperda* and *Diatraea lineolata*	Lower oviposition rates, trap cropping
Strip cropping of muskmelons with wheat	*Myzus persicae*	Interference with aphid dispersal
Oats intercropped with field beans	*Rhopalosiphum* sp.	Interference with aphid dispersal
Peaches intercropped with strawberries	Strawberry leafroller *(Ancylis comptana)* Oriental fruit moth *(Grapholita molesta)*	Population increase of parasites *(Macrocentrus ancyclivorus, Microbracon gelechise,* and *Lixophaga variabilis)*
Peanut intercropped with maize	Corn borer *(Ostrinia furnacalis)*	Abundance of spiders (*Lycosa* sp.)

TABLE 5.1 *(continued)*

Multiple-Cropping System	Pest(s) Regulated	Factor(s) Involved
Sesame intercropped with corn or sorghum	Webworms (*Antigostra* sp.)	Shading by the taller companion crop
Sesame intercropped with cotton	*Heliothis* spp.	Increase of beneficial insects and trap cropping
Soybean strip cropped with snap beans	*Epilachna varivestis*	Trap cropping
Squash intercropped with maize	*Acalymma thiemei, Diabrotica balteata*	Increased dispersion due to avoidance of host plants shaded by maize and interference with flight movements by maize stalks
Tomato and tobacco intercropped with cabbage	Flea beetles (*Phyllotreta cruciferae*)	Feeding inhibition by odors from nonhost plants
Tomato intercropped with cabbage	Diamondback moth (*Plutella xylostella*)	Chemical repellency or masking

Source: Based on Altieri et al. (1978), Altieri and Letourneau (1982), and Andow (1991a).

found when data were analyzed by parasitoid species or group of species. Only 54 percent of the thirty-one studied species had a higher parasitism rate or density in intercrops compared to monocultures (39 percent showed similar or variable activity levels in simple and diverse habitats). These data suggest that the response of some species to intercropping differs with different crop combinations, geographic location, and experimental procedures.

Experiments reporting results in which no differences were observed or in which higher pest incidence occurred in multicrops are quite uncommon. A particular crop mix might be of value in controlling one pest in one area (i.e., *Heliothis virescens* in corn *[Zea mays]* and cotton [*Gossypium* sp.] strip cropping in Peru), while increasing the same pest in other areas (i.e., *H. virescens* in Tanzania) (Smith and Reynolds, 1972). In Nigeria, populations of flower thrips *(Megalurothrips sjostedti)* were reduced by 42 percent on cowpea *(Vigna unguiculata)*-maize polycultures. However, cropping pattern had no effect on infestations of *Maruca testulalis,* pod-sucking bugs, and meloid beetles (Matteson, Altieri, and Gagne, 1984). Early infestations of *Maruca* were no different in monocrops and polycultures of maize and cowpea, but twelve weeks after planting, infestations were significantly higher in the monocrops. Similar shifts were observed with *Laspeyresia* and thrips (Matteson, Altieri, and Gagne, 1984).

In India, larval populations of *Heliothis armigera* were higher in sorghum *(Sorghum bicolor)*-pigeon pea *(Cajanus cajan)* intercropping systems than in sole pigeon pea plots, which led to higher grain losses in polycrops (Bhatnagar and Davies, 1981). In home-garden plots of beans *(Phaseolus vulgaris)* bordered by marigolds *(Tagetes* spp.), Latheef and Irwin (1980) reported that their designs did not favor control of *Heliothis zea* and *Epilachna varivestis.* In Georgia, Nordlund, Chalfant, and Lewis (1984) did not find significant reductions of *Heliothis zea* damage in maize ears, bean pods, or tomato fruits in polycultures of maize, bean, and tomato. In the Philippines, Hasse and Litsinger (1981) found that intercropping maize with legumes did not reduce the numbers of egg masses laid by corn borers *(Ostrinia furnacalis).*

Certain associated plants can have an adverse effect on parasitoids by acting as traps for searching adults. For example, *Lysiphlebus testaceipes* (Hymenoptera: Aphiidae), a parasitoid of cotton aphid *(Aphis gossypii)* (Homoptera: Aphidae), was entrapped by glandular exudates of petunia, and thereby prevented from protecting nearby okra *(Abelmoschus esculentus)* plants from aphid attacks (Marcovitch, 1935); *Trichogramma minutum* (Hymenoptera: Trichogrammatidae) adults could not parasitize hornworms *Manduca* sp. (Lepidoptera: Sphingidae) on tomato *(Lycopersicon esculentum)* plants because of entrapment by the sticky glandular trichromes on the leaves of the associated tobacco *(Nicotiana tobacum)* plants (Marcovitch, 1935). Although high parasitism of *Heliothis armigera* eggs by *Trichogramma* sp. occurred in sorghum *(Sorghum halepense),* a low parasitism rate was recorded in sorghum associated with chickpea *(Cicer arietinum)* because adult wasps were trapped in its sticky trichromes (Van Emden, 1990).

Despite these reports, use of intercropping has been widely recommended as a management strategy to reduce insect damage. A reduced insect-pest incidence in multicrops may be the result of increased parasitoid and predator populations, higher availability of alternate food for natural enemies, decreased colonization and reproduction of pests, chemical repellency, masking and/or feeding inhibition from nonhost plants, prevention of pest movement and/or emigration, and optimum synchrony between pests and natural enemies (Matteson, Altieri, and Gagne, 1984). Perrin and Phillips (1978) described the stages in pest population development and dynamics that may be affected by mixed cropping. At the crop colonization stage, they postulate that disruption of olfactory and visual responses, physical barriers, and diversion to other hosts are important mechanisms regulating herbivores in multiple-cropping systems. Once the pests become established in the field, their populations may be regulated by limitation of dispersal, feeding disruption, reproduction inhibition, and mortality imposed by biotic agents (Figure 5.1).

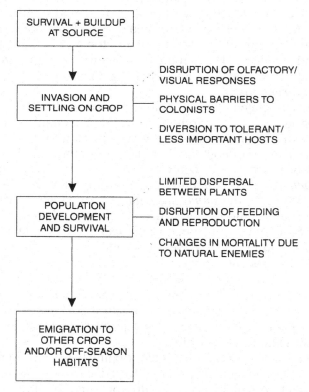

FIGURE 5.1. Stages in pest-population dynamics which may be affected by mixed cropping (after Perrin and Phillips, 1978).

Hasse and Litsinger (1981) summarized several mechanisms that supposedly explain pest reduction in intercropping systems. A list of the proposed mechanisms is given in Table 5.2. Most of the proposed mechanisms are accounted for by the resource concentration and enemies hypotheses discussed in Chapter 3.

HERBIVORE TRENDS IN POLYCULTURES

Andow (1991a) contends that herbivore movement patterns are more important than the activities of natural enemies in explaining the reduction of monophagous pest populations in diverse annual crop systems. Two classic studies support this view. The first study (Risch, 1981) looked at the population dynamics of six chrysomelid beetles in monocultures and polycultures of maize-bean-squash *(Cucurbita pepo)*. In polycultures containing at least one nonhost plant (maize), the number of beetles per unit was significantly lower relative to the numbers of beetles on host plants in monocultures. Measurement of beetle movements in the field showed that beetles tended to emigrate more often from polycultures than from host monocultures. Apparently, this was due to several factors: (1) beetles avoided host plants shaded by maize; (2) maize stalks interfered with flight movements of beetles; and (3) as beetles moved through polycultures, they remained on nonhost plants for a significantly shorter time than spent on host plants. There were no differences in rates of parasitism or predation of beetles between systems.

The second study examined the effects of plant diversity on the cucumber beetle, *Acalymma vittata* (Bach, 1980a). Population densities were significantly greater in cucumber *(Cucumis sativus)* monocultures than in polycultures containing cucumber and two nonhost species. Bach also found greater tenure time of beetles in monocultures than in polycultures. She determined that these differences were caused by plant diversity per se and not by differences in host-plant density or size. Thus, they do not reveal if differences in numbers of herbivores between monocultures and polycultures are due to diversity or rather to the interrelated and confounding effects of plant diversity, plant density, and host-plant patch size.

In northern California, densities of cabbage aphids *(Brevicoryne brassicae)* and flea beetles *(Phyllotreta cruciferae)* were signifi-

TABLE 5.2. Possible Effects of Intercropping on Insect-Pest Populations

Factor	Explanation	Example
Interference with host-seeking behavior		
Camouflage	A host plant may be protected from insect pests by the physical presence of other overlapping plants.	Camouflage of bean seedlings by standing rice stubble for beanfly
Crop background	Certain pests prefer a crop background of a particular color and/or texture.	Aphids, flea beetles, and *Pieris rapae* are more attracted to cole crops with a background of bare soil than to ones with a weedy background
Masking or dilution of attractant stimuli	Presence of nonhost plants can mask or dilute the attractant stimuli of host plants leading to a breakdown of orientation, feeding, and reproduction processes.	*Phyllotreta cruciferae* in collards
Repellent chemical stimuli	Aromatic odors of certain plants can disrupt host-finding behavior.	Grass borders repel leafhoppers in beans; populations of *Plutella xylostella* are repelled from cabbage-tomato intercrops
Interference with population development and survival		
Mechanical barriers	All companion crops may block the dispersal of herbivores across the polyculture. Restricted dispersal may also result from mixing resistant and susceptible cultivars of one crop by settling on nonhost components.	
Lack of arrestant stimuli	The presence of different host and nonhost plants in a field may affect colonization of herbivores. If a herbivore descends on a nonhost, it may leave the plot more quickly than if it descends on a host plant.	
Microclimatic influences	In an intercropping system, favorable aspects of mircroclimate conditions are highly fractioned; therefore, insects may experience difficulty in locating and remaining in suitable microhabitats. Shade derived from denser canopies may affect feeding of certain insects and/or increase relative humidity which may favor entomophagous fungi.	
Biotic influences	Crop mixtures may enhance natural-enemy complexes (see natural-enemy hypothesis in text).	

Source: After Hasse and Litsinger, 1981; and Litsinger et al., 1991.

cantly lower on cauliflower plants grown in association with vetch (*Vicia* sp.) than in clean-cultivated monocultures (Altieri, 1984). The depression of crop growth and biomass in the diverse plots added a confounding effect in that it was not clear if herbivore reduction resulted from poorer plant quality, which made cauliflowers less attractive to the herbivores.

In another study, flea beetle numbers were significantly lower in collards associated with wild mustard *(Brassica campestris)* than in monocultures (Altieri and Gliessman, 1983). Flea beetles preferred this plant over collards, thus flea beetles were diverted from collards resulting in diluted feeding on the collards. The authors argue that wild mustards have higher concentrations of allylisothiocyanate (a powerful attractant to flea beetle adults) than do collards, and therefore the preference of flea beetles for wild mustard simply reflected different degrees of attraction to the foliage levels of this particular glucosinolate in the weeds and collards. Figure 5.2 illustrates this

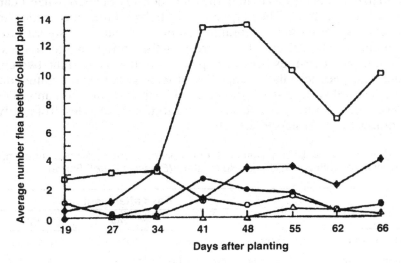

□ collard monoculture-normal density
♦ collard monoculture-double density
● wild mustard intercrop
△ barley intercrop

FIGURE 5.2. Populations trends of *Phyllotreta cruciferae* in collard monocultures and in collard polycultures mixed with a host plant (wild mustard) and a nonhost plant (barley) (after Altieri and Schmidt, 1986b).

preference in the field by showing that population densities of flea beetles on collard plants grown as monocultures are greater than on collards intercropped with wild mustards and with nonhost barley *(Hordeum vulgare)* (Altieri and Schmidt, 1986b). Although the barley effect might support the resource concentration hypothesis, the trap-cropping effect of wild mustards exerts a stronger influence on beetle abundance in this case.

The same study also showed that removal of flowers of wild mustards results in a substantial reduction of the attractant effect (Table 5.3). Consequently, collard plants in flowerless intercrops experienced greater flea beetle loads than collards within the intercrop with flowers and even monocultures.

Risch, Andow, and Altieri (1983) collected additional data that do not support the enemies hypothesis. They found that predation rates on egg masses of the European corn borer *(Ostrinia nubilalis)* by a predaceous beetle *(Coleomegilla maculata)* were significantly higher in maize monocultures than in the more densely planted maize-bean-squash polyculture. They argue that in polycultures, the beetles apparently spent more time foraging on plants (beans and squash) that contained no borer eggs, thus decreasing their foraging effectiveness. Even if prey densities per maize plant were the same in the two culture types, beetles might forage less efficiently in the polyculture due to unrewarded time spent foraging on bean and squash plants. This lower reward rate leads to faster emigration of beetles from polycultures (Wetzler and Risch, 1984).

TABLE 5.3. Flea Beetle Numbers per Collard Plant in Monocultures and Polycultures with and Without Wild Mustard *(Brassica kaber)* Flowers

		Days after planting*		
		30	**44**	**57**
Monoculture:	normal density	30.0 ± 6.9	49.6 ± 19.7	10.1 ± 6.6
Monoculture:	double density	40.0 ± 24.0	79.6 ± 43.6	6.5 ± 2.8
Polyculture:	with flowers	5.6 ± 2.5	10.6 ± 5.5	1.6 ± 0.6
Polyculture:	without flowers	**7.6 ± 3.5	81.0 ± 38.3	5.8 ± 2.4

Source: After Altieri and Schmidt, 1986b.

*Means derived from different counts on 10 random plants per plot
**Flowers not yet removed

Wrubel (1984) contends that visual camouflage from nonhost plants may have resulted in more Mexican bean beetles colonizing soybean *(Glycine max)* monoculture than maize-soybean intercropped plots. Conversely, the higher concentration of food resources in clover-soybean (two legumes) than soybean monoculture plots may explain the slightly higher abundance of polyphagous acridids that Wrubel found in the clover-soybean plots. Differences in the structure of the crop canopy in tall maize-soybean and short maize-soybean plots appeared to affect the behavior of several groups of herbivores, with lower abundance of Japanese beetles due to shading of the soybean canopy by the taller maize plants.

There are, however, studies that support the enemies hypothesis. In tropical corn-bean-squash systems, Letourneau (1983) studied the importance of parasitic wasps in mediating the differences in pest abundance between simple and complex crop arrangements. A squash-feeding caterpillar, *Diaphania hyalinata* (Lepidoptera: Pyralidae), occurred at low densities on intercropped squash in tropical Mexico. Part of the effect of the associated maize and bean plants may have been to render the squash plants less apparent to ovipositing moths. Polyculture fields also harbored greater numbers of parasitic wasps than did squash monocultures. Malaise trap captures of parasitic wasps in monoculture consisted of one-half the number of individuals caught in mixed culture. The parasitoids of the target caterpillars were also represented by higher numbers in polycultures throughout the season. Not only were parasitoids more common in the vegetationally diverse, traditional system, but also the parasitization rates of *D. hyalinata* eggs and larvae on squash were higher in polycultures. Approximately 33 percent of the eggs in polyculture samples over the season were parasitized while only 11 percent of eggs in monocultures were. Larval samples from polycultures showed an incidence of 59 percent parasitization for *D. hyalinata* larvae, whereas samples from monoculture larval specimens were 29 percent parasitized.

Another study conducted in Davis, California, tested whether predator colonization rates could be manipulated through vegetational diversity (Letourneau and Altieri, 1983). The densities of *Orius tristicolor* and its preferred prey, *Frankliniella occidentalis,* were compared between squash monocultures and polycultures of squash, corn, and cowpea. The patterns of predator colonization rates and pest densities in these two cropping systems paralleled those documented for predator-prey interactions in the mite-grapevine systems of Flaherty (1969). In both studies, the colonization rate of predators was increased in diverse habitats, and the prey (pest) populations in each case reached lower maximum levels. In Flaherty's study, the causative factor was the close proximity of the source of colonizing predators. The great variation between levels of Williamette mite *(Eotetranychus willamette)* infestation on individual grape vines was caused by their variable proximity to clumps of johnsongrass *(Sorghum halepense)*. The grass supported an alternate host for a predatory mite, *Metaseiulus occidentalis.* The predators then colonized contiguous vines sufficiently early to suppress the pest-mite populations. In the Davis study, the sources of colonizers were presumably at similar distances to randomly assigned plots of monoculture and polyculture. The authors suggest that the determining factor for differential colonization by *Orius* sp. in monocultures and polycultures of squash was attraction to the early-season complex-crop habitat during the host-location process. Results showed that mean density of thrips on squash leaves was initially much greater in monoculture than in polyculture and remained at significantly higher levels until sixty-five days after sowing. The *Orius* sp. density, however, was significantly higher on squash early in the season (days 30 and 42) in polyculture. A decrease in prey density accompanied an increase in adult *Orius* sp. colonization in both treatments until thrips reached low densities (Figure 5.3).

Predator manipulation experiments conducted in field cages, in which *Orius* sp. populations were either included or excluded, showed that the density of thrips was influenced by predation by *Orius* sp. (Letourneau and Altieri, 1983). On uncaged control plants, the mean density of thrips per leaf declined steadily from day 50, as it had in the general field samples. Inside the exclusion/inclusion cages, thrip density more than tripled the first week after *Orius* and *Erigone* spp. spiders were eliminated. When predators, equal in number to those that were eliminated, were added to cage 1, the thrip density fell in this cage.

Altieri (1984) found that brussels sprouts grown in polycultures with fava beans or wild mustard supported more species of natural enemies (six species of predators and eight species of parasites) than monocultures (three species of predators and three species of parasites). Apparently, the presence of flowers, extrafloral nectaries, and

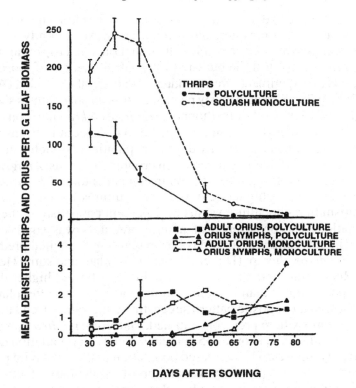

FIGURE 5.3. Predator-prey relationships in monoculture and polyculture plots from samples taken on squash leaves (after Letourneau and Altieri, 1983).

alternate prey/hosts associated with the companion plants allowed this enhancement. Aphid densities were also lower in such systems, apparently due to increased mortality imposed by the more diverse species complex of natural enemies.

CASE STUDY 1: MAIZE INTERCROPS AND PEST ATTACK

Corn-bean polycultures are common diversified systems used by small farmers in Latin America. These intercropping systems usually result in higher yields for several reasons, including reduced weed competition due to dense crop cover, soil conservation, and better use of incident radiation, water, and soil nutrients. Another advantage is that insect-pest attack is often less than that on sole crops (van Huis, 1981). In order to test the built-in pest-control features of those systems, several experiments were conducted in tropical Colombia (Altieri and Doll, 1978). Simultaneously planted corn-bean polycultures showed significantly fewer adult leafhoppers *(Empoasca kraemeri)* on beans in the maize-bean polyculture compared to monoculture beans, until the final sampling date seventy days after planting. Nymphal populations were not affected by diversity in cropping systems. *Anagrus* sp. (Hymenoptera: Mymaridae), the main egg parasitoid of *E. kraemeri,* showed 20 percent higher activity in polycultures with 48.5 percent parasitism in monoculture versus 60.7 percent parasitism in the crop association. The occurrence of natural predators was significantly higher in polyculture at forty days from planting. Principal predators were *Condylostylus* sp. (Diptera: Dolichopodidae) and some Hemiptera (Reduviidae and Nabidae). These insects showed higher densities in polycultures than in monocultures, but fifty days after planting they showed an opposite pattern suggesting a migratory trend toward monocultures where prey concentrated. *Diabrotica balteata,* a polyphagous leaf beetle, showed 45 percent lower adult population densities in polycultures. The increased population of their reduviid predators in polycultures probably was an important regulatory factor. It is possible that the presence of other chrysomelids, which was 30 percent higher in polyculture, exerted a competitive displacement of *D. balteata,* decreasing its feeding and colonization efficiency. The number of bean plants with damaged leaves was similar in polycultures and monocultures.

In general, the percentage of corn seedlings damaged by cutworm attack was low in all treatment plots; however, it tended to be lower in plots with high vegetational diversity. Corn-bean polycultures had the lowest number of damaged seedlings. Twenty days after planting, fall armyworm *(Spodoptera frugiperda)* (FAW) larval densities were significantly lower in polycultures than in monocultures at all sampling dates. Parasitism of FAW larvae by the braconid *Meteorus* sp. forty days after planting was higher in the polycultures than in the monocultures.

Intercropping corn with different bean varieties had diverse effects upon FAW. Corn interplanted with the bush bean variety ICA Pijao presented 14 percent less whorl damage than corn interplanted with the climber bean variety P-589. Systems with sequential planting of corn in relation to beans resulted in reduced population densities of adult *E. kraemeri* when maize was planted twenty to forty days before the beans. Significantly lower infestation levels of FAW in corn were observed with beans planted twenty to thirty days before, and higher and uniform populations in treatments where beans were planted ten days before to twenty days after corn. The FAW populations in corn were reduced by 88 percent by the early plantings of beans. Corn yields did not vary significantly among these planting dates.

In surveys of maize-bean intercrops in Nicaragua, van Huis (1981) found trends of FAW similar to those found in Colombia. In addition to lower incidence of FAW, he noted population reductions of the stalk borer *Diatraea lineolata.* In Mexico, Trujillo-Arriaga and Altieri (1990) found that maize associated with fava bean and squash exhibited lower damage by *Tetranychus urticae* than monocultures, apparently because maize in monocultures was more affected by water stress, a condition that makes these plants more susceptible to mite attack. In polycultures, the aphid *Rhopalosiphum maidis* experienced higher attack by several species of lady beetles, *Hippodamia convergens* and *H. koebelei.*

A corn-diversification strategy amenable to U.S. midwestern farmers is strip cropping of corn and soybeans. Tonhasca (1993) reported a general positive response to strip-cropping systems by several predator species (including lady beetles, spiders, minute pirate bugs, and nabids) and by parasitic wasps. Corn in strip-cropping plots provided shade, reduced wind speed, higher humidity, lower temperatures, and alternate food, which favored all factors for soybean natural enemies.

CASE STUDY 2: CASSAVA INTERCROPS AND PEST INCIDENCE

Traditional farmers in Latin America and Africa commonly intercrop cassava, partly as a defense against herbivore attack. For example, in Nigeria, farmers intercropping cassava with maize and sorghum reported lower populations of *Zonocerus variegatus* than in monocultures. In Colombia, cassava intercropped with beans exhibits reduced populations of various pests such as the hornworm, shoot fly, and lacebug. Cassava whitefly species (*Aleurotrachelus socialis* and *Trialeurodes variabilis*) exhibited lower per-leaf densities in cassava-cowpea intercrops than in monoculture. Intercropping with maize had no significant effect on whitefly populations per leaf (Gold, Altieri, and Bellotti, 1989a). Rates of parasitism and overall immature mortality of cassava whiteflies were similar among mono- and polyculture.

In order to determine yield losses under the different cropping systems and to separate the effects of intercrop competition and differential herbivore numbers, applications of the insecticide monocrotophos were made to protected plots. Intercropping with cowpea lowered yields of a regional cultivar of cassava ('MCOL 2257') in protected plots. However, in nonprotected plots, the regional cassava variety intercropped with cowpea had higher yields and sustained lower yield losses than other systems. Yields of regional cassava intercropped with maize, grown in monoculture, or mixed with cassava cultivar CMC 40 were equivalent in both protected and nonprotected environments. Yield losses closely followed population trends of cassava whiteflies. Whiteflies were attracted to more vigorous plant assemblages, as in monocultures, with lowest numbers in cassava-cowpea systems. However, the data indicate that under stress, cassava favors top growth over roots, and large plant size did not ensure high yield. Land equivalent ratios exceeded 1.5 for intercropped systems (Gold, Altieri, and Bellotti, 1989b).

One of the intriguing aspects of the cassava-cowpea association is that reductions in whitefly densities persist even after the harvest of cowpea (at three months, when cassava stays in the field an additional six to nine months). Both species of whitefly (*A. socialis* and *T. variabilis*) had lower egg densities on cassava-cowpea mixtures than on cassava monoculture, with lower levels remaining for six months. These residual effects apparently resulted from two possible sources. The existent populations of herbivores and their natural enemies at the time of intercrop harvest may have influenced pest population dynamics during the postintercrop period. The relative mobility of the insects and the determination of immigration and emigration rates may indicate how long such a residual affect may persist. In addition, differences in plant quality may extend far into the postintercrop period. Apparently, intercrop competition retarded cassava growth, causing reductions in host-plant size well beyond the intercrop period. Greater whitefly numbers were associated with more vigorous plants and, hence, were higher in monoculture (Gold, Altieri, and Bellotti, 1990).

CASE STUDY 3:
REDUCING STEMBORERS IN AFRICA

Many intercropping studies have transcended the research phase and have found applicability to control specific pests such as the stemborers in Africa. Scientists at ICIPE developed a habitat management system that uses two kinds of crops planted together with maize: a plant that repels these borers (the push) and another that attracts (pulls) its natural enemies (Khan et al., 2000). The push-pull system has been tested on over 450 farms in two districts of Kenya and has now been released for uptake by the national extension systems in East Africa. Participating farmers in the breadbasket of Trans Nzoia have reported a 15 to 20 percent increase in maize yield. In the semiarid Suba district, plagued by both stemborers and striga, a substantial increase in milk yield has occurred in the past years, with farmers now being able to support cows on the fodder produced. When farmers plant maize, napier grass, and desmodium together, a return of US$2.30 for every dollar invested is made, as compared to $1.40 obtained by planting maize as a monocrop. Two of the most useful trap crops that pull in the borers' natural enemies are napier grass *(Pennisetum purpureum)* and Sudan grass *(Sorghum bicolor Sudanese)*, both important fodder plants; these are planted in a border around the maize. Two excellent borer-repelling crops that are planted between the rows of maize are molasses grass *(Mellinis minutifolia)*, which also repels ticks, and the leguminous silverleaf maize *(Desmodium)*, which can suppress the parasitic weed *Striga* by a factor of forty compared to maize monocrop; its N-fixing ability increases soil fertility, and it is an excellent forage. As an added bonus, sale of desmodium seed is proving to be a new income-generating opportunity for women in the project areas.

LIVING MULCHES: A SPECIAL TYPE OF INTERCROP

The use of legume cover crops as sod-strip intercropping and/or living mulches in year-round cropping systems and rotations has been proposed as holding potential for sustained crop production and self-sufficiency in soil nutrients (Vrabel, Minnotti, and Sweet, 1980; Palada et al., 1983). Researchers at Cornell University and at the Rodale Research Center have found that overseeding legumes in corn and other annual crops maintains yields while providing increased soil protection on highly erosive soils. Moreover, when the legume sod is properly managed, weed suppression is enhanced significantly, reducing the need for chemical herbicides.

Although the entomological advantages of these systems are still poorly understood, experimental work suggests that many living-mulch systems have built-in biological control advantages. Most research has focused on *Brassica* crops. For example, Dempster and Coaker (1974) found that the maintenance of a clover cover aided in the reduction of three insect pests (*Brevicoryne brassicae, Pieris rapae,* and *Erioischia brassicae*). In the case of *P. rapae,* the reduction was attributable to increased numbers of the predacious ground beetle *Harpalus rufipes* in the clover-seeded plots. Similar enhancements were observed when planting clover between rows of cabbages, which resulted in a 34 percent increased predation of eggs of the cabbage root fly, *Delia brassicae* (Cromartie, 1981). Hooks, Valenzuela, and Defrank (1998) report decreased pest incidence due to enhanced natural enemy activity in zucchini grown with living mulches.

In New York State, an experiment was conducted using cabbage interplanted with several living mulches and in bare-ground monocultures (Andow et al., 1986). Living mulches included creeping bent grass, red fescue, Kentucky bluegrass, and two white clovers. Populations of *Phyllotreta cruciferae* and *Brevicoryne brassicae* were lower on cabbage grown with any living mulch than on cabbage in bare-ground monocultures. First-generation larvae of *Pieris rapae* were more common on cabbage with clover living mulches, but second-generation eggs and larvae were less common on cabbage with clover living mulches. These differences in population density were probably determined by variation in herbivore colonization rates, not by variation in herbivore mortality. The authors suggested that early- season chemical treatments for flea beetles might be eliminated when living mulches are used. However, this potential gain may be offset by yield reduction from competition between cabbage and living mulches.

Helenius (1998) reports a study of cabbage with or without clover in the interrow spaces in which cabbage root fly eggs were consistently reduced by 25 to 64 percent in the intersown plots. Laboratory experiments suggested that intersowing had not deterred oviposition. Similarly, pitfall traps did not provide evidence of higher numbers of epigeal predators. However, predator-exclusion experiments revealed that oviposition was indeed reduced (by 18 percent) in cabbages with clover and that predators further reduced egg numbers to 41 percent of control values.

In two locations in California, Altieri, Wilson, and Schmidt (1985) further tested the effects of vegetation background in the form of living mulches and natural weed cover on the population dynamics of foliage and soil arthropods in corn, tomato, and cauliflower crop systems. In Davis, California (Central Valley site), herbivores (especially aphids and lygaeids) were more abundant in the plots with weed cover than in the clover-mulch plots, whereas leafhoppers were most common in the clover mulch. Higher numbers of natural enemies were observed in the clover plots. Significantly more ground predators (Carabidae, Staphylinidae, spiders) were caught in pitfalls placed in the weedy and clover plots than were caught in the clean-cultivated plots. In Albany, California (coastal area), specialized herbivore (cabbage aphid and flea beetle) densities were significantly reduced in plots with living-mulch cover. It is not clear if this reduction was due to plant diversity or density effects, the effects of natural enemies, or the lower quality of plants in the weedy and mulched plots, as crop growth and yields were drastically reduced in these plots at both sites. In Salinas, California, Costello and Altieri (1994) designed an experiment to test whether a living mulch of white clover, strawberry clover, and bird's-foot trefoil, could be useful in place of insecticides to protect broccoli from cabbage aphid infestation. The study showed that living mulches can reduce cabbage aphid densities in harvested broccoli heads relative to broccoli grown without a cover crop. Differences in infestation levels between living mulches and clean cultivation are best explained by differences in aphid colonization rates. The correlation between light intensity and numbers of alates suggests that the lower-intensity light reflected from broccoli grown with living mulches is less attractive to incoming aphids than the higher-intensity light reflected from clean-cultivated broccoli. The strong association between numbers of alates and intensity in the yellow wavelength is interesting given that the cabbage aphid is known to be attracted to the color yellow (Costello and Altieri, 1995). The authors suggested that further agronomic work is warranted to minimize the competitive effects of legume covers on crops, so that the observed entomological advantages can be used in a practical way.

Planting lucerne *(Medicago littoralis)* in the same row as carrot significantly reduced the damage levels of the carrot rust fly. Although the abundance of generalist predators was higher in the lucerne-carrot plots than in monocultures, this enhancement did not explain the observed decreased carrot fly rust damage. Emigration of carrot rust flies from the intercropping system was higher than from monocultures, thus flies exhibited longer tenure times in single habitats (Cromartie, 1981).

In England, undersowing cereals with grass species (i.e., ryegrass) increases the activity and abundance of natural enemies. Potatoes undersown with perennial ryegrass have been found to have aphid populations reduced by up to 66 percent compared with plants in bare ground. Since the number of colonizing aphids did not appear to be affected by undersowing, increased mortality from natural enemies was suspected. The practice of undersowing appears to be one of the most effective means of enhancing aphid parasitism by *Aphidius* spp. in cereals (Burn, 1987). A similar effect was shown in Germany, where parasitism of *Metopolophium dirhodum* by two parasitoids was higher in wheat undersown with clover than in wheat monoculture (El Titi, 1986). In an experiment in England, ryegrass undersown in wheat was purposefully infested with *Myzus festucae* followed by a release of the parasitoid *Aphidius rhopalosiphi*. Thus, a parasitoid population was established on the ryegrass before *Sitobion avenae* invaded the wheat in the spring (Powell, 1986). Populations of the pest aphid on wheat were smallest on those plots which had developed the largest *M. festucae* populations in the spring.

METHODOLOGIES TO STUDY INSECT DYNAMICS
IN POLYCULTURES

Many methodological approaches have been tested in studies of monocultures and polycultures to explain the ecological mechanisms underlying the entomological effects of diversity. Risch (1981) studied beetle-movement behavior to see if this could account for lower numbers of beetles in maize-bean intercrops. He placed directional malaise insect traps on each side of every plot. When beetles flew out of the plot, some of them landed on the vertical trap walls and were caught in the collecting jars. By counting these trapped beetles and estimating the total number of beetles in the plot at that time from direct counts, he calculated a ratio of the two groups and called the ratio "tendency to emigrate," which measures the beetle's relative tendency to leave a plot once it has arrived. After sixty to sixty-five days, there was a much greater tendency to emigrate from the bean monoculture than from the bean polyculture. This corresponds with the observation that there were far fewer beetles on beans planted with maize in the polycultures than in the bean monocultures and that this large difference became apparent approximately sixty-five days after planting. Maize has an inhibitory effect on the presence of this insect species.

How does maize exert its inhibitory effect? Beans grown with maize are shaded more than beans in monocultures. One possibility is that the beetles avoid feeding in shaded areas, preferring to feed on plants that are not shaded. This was tested directly by constructing two large shade screens and suspending them 80 cm above the ground. One screen provided little shade, allowing 65 percent light transmission, and the other provided much more shade, allowing only 25 percent light transmission. Squash and bean plants were grown in the greenhouse and placed under these screens. Then the numbers of colonizing beetles on the plants were counted over a series of days. The results showed that there were always significantly more beetles under the light-shade screen than the dark.

Yet shade might not be the only way that the presence of maize interferes with beetle-flight behavior. To determine if a vertical obstruction, such as a maize stalk, could discourage beetle colonization in other ways, dry maize stalks were staked among potted bean plants, and a light screen was erected over the plants. Potted beans without maize stalks were also placed in a nearby area with a darker screen over them, so that the total amount of light reaching the plants in both areas was identical. Consistently, many more beetles were found in the beans without maize stalks, indicating that maize physically inhibited beetle colonization in ways other than by just increasing the overall shade of a microhabitat.

Although these experiments provided an indication of the underlying causes of the reduction in beetle numbers in maize-bean polycultures, they did not help in predicting numbers of beetles in different variations of the entire maize-bean-squash system. Risch (1980) also studied the influence of size of the plot and relative proportions of maize, beans, and squash on the number of beetles in the field. He observed and modeled the movement of one beetle, *Acalymma vittata,* a squash specialist that is much more abundant in monocultures of squash than in maize-bean-squash polycultures. The variables Risch thought might be important in ultimately determining the rate at which a beetle leaves a maize-bean-squash polyculture versus a squash monoculture are the following: the time a beetle spends on a maize, bean, or squash plant; the probability of moving to a maize, bean, or squash plant; the distance a beetle travels when it leaves a maize, bean, or squash plant and flies over an intercrop or monoculture; and its orientation behavior at the edge of a plot.

Kareiva (1983) proposed that emigration models can be mathematically formalized as continuous-time, finite-state Markov processes in which insects move among three states (on host plant, on nonhost plant, outside plot) in polycultures but only between two states (on host plant, outside plot) in monocultures. Because computations and parameter estimation are simple for these models, this approach is ideal for exploring the connection between herbivory and trivial movement. Instantaneous transition rates between these states can be easily obtained by releasing and recapturing marked insects. Equilibrium herbivore densities for polycultures versus monocultures can then be calculated from the model; with this approach, it might be possible to explicitly attribute reduced pest pressure in polycultures to high rates of movement, either from host to nonhost plants or from nonhost plants to areas outside the crop.

One of the suggestions of Kareiva (1983) is that nonrandom pest movement is the process often responsible for the different damage levels observed in randomized-block field experiments. Individual insects of these mobile species will move frequently among the various cropping treatments included in a randomized block design; typically they will spend disproportionate amounts of time in the treatments that represent a preferred food or preferred habitat.

To evaluate the dependence of apparent crop protection on local pest movement, Kareiva established one set of treatment plots in close proximity to one another with free access between them, and a second set of treatment plots (also in close proximity) separated from one another by tall (1.5 m) curtain barriers, which interfere greatly with flea beetle movement (Figure 5.4). The cropping treatments that he contrasted were (1) pure stands of collards versus collards inter-

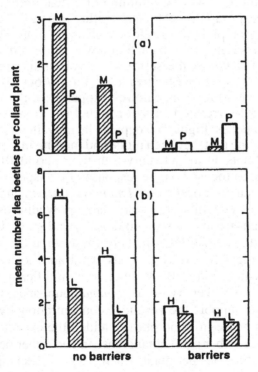

FIGURE 5.4. Effects of between-treatment movements on the response of flea beetles to cropping systems (polyculture [P] versus monoculture [M] {a}, or low density [L] versus high density [H] {b}). Movement between the top two subplots in each block was restricted by a curtain barrier. These cultivated blocks were surrounded by old-field vegetation constituted mainly of goldenrod (*Solidago* spp.) (after Kareiva, 1986).

cropped with potatoes and (2) collards planted at a high density of $6.7/m^2$ versus collards planted at a lower density of $3.3/m^2$. Where there were no barriers between intercropped and monoculture treatments, flea beetles were less abundant in the intercropped plots than in adjacent monocultures; where there were barriers between the treatments, intercropping yielded no reduction in beetle infestation (Kareiva, 1986).

By marking beetles in California, Garcia and Altieri (1992) examined the movement behavior and the rate at which marked flea beetles released in broccoli monocultures and broccoli-*Vicia* polycultures tended to either leave or stay in the system, or even migrate from one system to the other. After vacuuming of all naturally occurring flea beetles, three groups of 350 flea beetles marked with fluorescent blue, orange, or pink were released in each plot. The number of marked beetles remaining in each plot was estimated visually by inspecting all plants and the surrounding soil at six and twenty-four hours after release. More beetles tended to fly out and leave mixed cultures compared to monocultures. Apparently, the broccoli-*Vicia* system had a deterrent effect that resulted in massive emigration of the released beetles. Figure 5.5 depicts the movement patterns of flea beetles out of the plot, between plots, and moving into plots from surrounding habitats during a twenty-four-hour period after release.

Bach (1980a) focused on the response of one specialist herbivore, the striped cucumber beetle *(Acalymma vittata),* to cucumber monocultures versus cucumber-broccoli-maize polycultures. By controlling total plant density, host-plant density, and plant diversity, Bach was able to distinguish the effects of these three confounding variables. Applying a three-way analysis of variance to censuses of beetles per cucumber plant, Bach reported a significant effect of both plant density and diversity on *Acalymma* abundance, but the results only partially support the resource concentration hypothesis. Although an increase in stand purity yielded the expected increase in beetles per cucumber plant, an increase in cucumber density reduced the number of beetles per plant (Bach, 1980a). Bach also found that cucumber plants were, on average, smaller in polycultures than in monocultures and that beetle density was positively correlated with plant size. Two polyculture plots with cucumbers were equal in size to monoculture cucumbers; in these two thriving polycultures, beetles were still significantly fewer per plant than in monocultures.

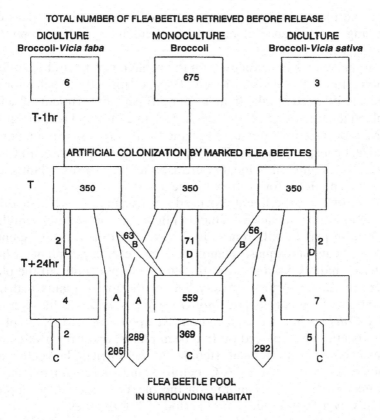

TOTAL NUMBER OF FLEA BEETLES RETRIEVED BEFORE RELEASE

FIGURE 5.5. Flux of marked flea beetles in three different cropping systems during a twenty-four-hour period after release. A = beetles emigrating out of crop habitat; B = beetles moving from dicultures to monoculture; C = beetles colonizing plots from surrounding habitats; D = beetles staying in the plots (after Garcia and Altieri, 1992).

Thus, it is clear that the reduced beetle numbers in polycultures cannot simply be attributed to smaller cucumber plants. In a later study with *Acalymma vittata,* Bach (1980b) provided evidence for a surprising reduction in foliage palatability associated with cucumber-tomato polycultures. Beetles given a laboratory choice between cucumber leaves grown in monoculture and cucumber leaves grown in a tomato-cucumber mixture significantly preferred monoculture leaves. This illustrates the subtle links that are possible between plant diversity and plant quality, quite apart from the conventional ideas concerning the influence of resource concentration on herbivores (Kareiva, 1983).

Some studies have focused on the behavior of natural enemies in polycultures. Wetzler and Risch (1984) examined the behavior of a coccinellid beetle in the field in four diffusion experiments, Each involved the release of beetle populations in a matched pair of agricultural plots (10 m × 10 m each) planted with various combinations of maize, beans, and squash. The day preceding each release, all *Coleomegilla maculata* individuals were aspirated from every plant, ensuring "clean" fields for each experiment.

One of the experiments focused on determining whether differences in diffusion rates from monocultures and polycultures might be caused in part by differences in the average time a beetle spent on maize, bean, and squash plants (i.e., tenure time per plant). Maize, squash, and bean plants were first grown in pots until all the plants were in flower. Approximately half of the maize plants had large numbers of the corn aphid *Rhopalosiphum maidis*. In the first trial, fifty *Coleomegilla* were placed on five aphid-infested maize plants, fifty beetles were placed on five bean plants, and fifty beetles were placed on five squash plants (ten beetles per plant). The beetles were cooled to approximately 6°C before being placed on plants. The number of beetles remaining on the plants was counted approximately every ten minutes for a period of 100 minutes.

Sight counting proved to be an effective means of population censusing since the beetles are highly visible, thus avoiding problems associated with trapping. Careful collection of individuals for release enabled uniform, almost equivalent releases of adults. Since each experiment was run for only twenty-four hours and was preceded by a minimum of handling of beetles, mortality was extremely low (0.5 percent) and complications due to beetle reproduction were nonexistent. The timing of the one-hour, three-hour, and six-hour censuses was arranged to correspond with maximum periods of diurnal activity to ensure that the most conservative diffusion estimates would arise during the final censuses. Since all experiments were conducted within a five-week interval, seasonal variability (i.e., migratory movements) of *Coleomegilla* activity was restricted.

In their studies of corn-cowpea-squash polycultures, Letourneau and Altieri (1983) found visual-inspection sampling of thrips and *Orius* to produce a more representative measure of density than did sticky traps, pan traps, or malaise traps, each of which showed very few catches. Ten hills of squash (each hill consisting of two plants) were randomly selected, and the plant most southwest in the hill was sampled by gently turning each leaf and recording the numbers of *Orius* adults and nymphs (as well as any other common arthropods). Thrips were counted on one medium-sized leaf of each plant. During this season, *Orius* densities increased, and plants grew so large that the number of leaves sampled was reduced to five per plant: the growing shoot, two young, and two old leaves. Biomass estimates were made at two-week intervals by measuring leaf widths on all the leaves of ten plants per plot. The leaf width of squash plants was highly correlated with leaf biomass, determined as dry weight of the leaf blade ($r = 0.93$). To standardize for possible leaf size differences between treatments (and thus searching-area differences), predator numbers per plant were converted to numbers per 5 g of leaf biomass. Individual leaves sampled for thrips were also measured to allow for conversion of thrips per leaf to thrips per 5 g of leaf biomass.

To determine whether predators were concentrated within treatments on plants with higher thrip densities, an index of aggregation was calculated on day 30. Ratios of mean thrip density on plants with *Orius* to those without *Orius* present would be significantly greater than if *Orius* were showing such a preference within a treatment.

Coll (1998) has warned about the limitations of using trapping or recording host-parasitism rates in simple and diverse habitats. These methods are appropriate only when habitat type does not affect the precision of the sampling. However, sampling may not be equally effective in different habitats because differences in plant height and architecture may alter sampling effort and/or effectiveness. For example, Perfect (1991) used traps to compare parasitoid density when cowpea and maize were grown alone or intercropped. They found that when traps were placed 0.5 m above the ground, almost twice as many chalcidoids were trapped in the intercrop than in monocultures. However, when the traps were situated 2 m above the ground, similar numbers of chalcidoids were trapped in the two monocultures and in the intercrop. Similarly, maize height affected the number of tachinid flies caught in malaise traps. Thus, in both cases the traps are inappropriate for comparing parasitoid abundance in monocultural and intercropped habitats. Colored traps may be more apparent in one habitat, resulting in greater attractiveness and capture of parasitoids. Estimating parasitism rate may also depend on habitat type if, for example, host spatial distribution (i.e., clumpedness) differs among habitats but the same sampling protocol is used in different habitats.

Given these limitations, Coll and Botrell (1996) used release-recapture experiments to determine how the presence of maize (non-host plant) influenced the movement of the parasitoid *(Pediobius foveolatus)* in bean plots in the absence of hosts. They tested three predictions that are based on the response of monophagous herbivores to plant diversification. In the absence of hosts they tested whether (1) the parasitoid immigrates more readily to taxonomically simple than diverse habitats, (2) it is more likely to remain in simple than in diverse plant stands, and (3) its movement in the habitat is hampered by the presence of tall nonhost plants. Then they assessed how vegetation diversity affects wasp reproduction (parasitism) and subsequent density in the presence of its hosts, Mexican bean beetle larvae.

Fewer female wasps immigrated into and more emigrated out of a bean and tall maize intercrop than bean monocultures. Bean plant density and the presence of maize per se did not significantly affect parasitoid immigration. Instead, maize height was the primary factor contributing to lower female immigration into the intercropped bean and tall maize plots. However, tall maize plants did not impede the wasps' within-habitat movement.

When wasps were released outside the plots, higher parasitism was recorded in monocultures, irrespective of host density. In contrast, when wasps were released within the plots, significantly higher parasitism rates were found in the bean and tall maize habitat. Results suggested that female wasps accumulate in the bean and tall maize habitat in response to resources other than hosts and, ultimately, wasp density may be determined primarily by differential emigration rather than by immigration rates.

These research examples illustrate the range of methodologies that have been employed to explain insect movement and differences in densities in simple and complex habitats. These studies have been crucial in advancing our understanding of how plant diversity influences insect ecology in polycultures.

MANAGEMENT CONSIDERATIONS

Multiple-crop management is basically the design of spatial and temporal combinations of crops in an area (Harwood, 1979). There are many possible crop combinations and arrangements, and each can have different effects on insect populations. The attractiveness of crop habitats to insects in terms of size of field, nature of surrounding vegetation, plant densities, height, background color and texture, crop diversity, and weediness is subject to manipulation. An important goal of intercropping research should be to fully understand the mechanisms involved in pest reduction in polycultures, so that informed manipulations can be done to improve the entomological advantages of intercropping systems.

In intercrop systems, the choice of a tall or short, early or late maturing, flowering or nonflowering companion crop can magnify or decrease the effects on particular pests (Altieri and Letourneau, 1982). The inclusion of a crop that bears flowers during most of the growing season can condition the buildup of parasitoids, thus improving biological control. Similarly, the inclusion of legumes or other plants supporting populations of aphids and other soft-bodied insects that serve as alternate prey/hosts can improve survival and reproduction of beneficial insects in agroecosystems. Early planting of an aphid-supporting legume in patches within the field can initiate the buildup of parasitoids before the rest of the crop is planted and more aphids colonize the field. The presence of a tall associated crop such as maize or sorghum may serve as a physical barrier or trap to pests invading from outside the field.

Tall plants can affect the visual stimuli by which insect pests orient themselves to their suitable host plants or may interfere with the herbivore's movement and dispersal within the system (Perrin, 1977). The inclusion of strongly aromatic plants such as onion *(Allium cepa),* garlic *(Allium sativum),* or tomato *(Lycopersicon esculentum)* can disturb mechanisms of orientation to host plants by several pests. Including onions as intercrops in carrot fields in England reduced the attack of carrot fly *(Psila rosae)* and carrot willow aphid *(Cavanella aegopodii)* (Uvah and Coaker, 1984).

The date of planting of component crops in relation to one another can also affect insect interactions in these systems. An associated crop can be planted so that it is at its most attractive growth stage at the time of pest immigration or dispersal, diverting pests from other more susceptible or valuable crops in the mixture. Planting of okra *(Hibiscus esculentus)* to divert flea beetles *(Podagria* spp.) from cotton in Nigeria is a good example (Perrin, 1980). Maize planted thirty and twenty days earlier than beans reduced leafhopper population on beans by 66 percent compared to populations in plots under simultaneous planting. Fall armyworm damage on maize was reduced by 88 percent when beans were planted twenty to forty days earlier than maize when compared to simultaneous planting (Altieri, Schoonhoven, and Doll, 1977).

We still understand little how spatial arrangements (e.g., row spacings) of crops affect pest abundance in intercrops. For example, there is greater reduction in damage to cowpea flowers by *Maruca testulalis* in intrarow rather than interrow mixtures of maize and cowpea. Selection of proper crop varieties can also magnify insect suppression effects. In Colombia, lower whorl damage by *Spodoptera frugiperda* was observed in maize associated with bush beans than in maize mixed with climbing beans. In the same trials, maize hybrid H-207 seemed to exhibit lower *Spodoptera* damage than hybrid H-210 when intercropped with beans (Altieri et al., 1978). Strip-cropping systems can preferentially act as trap crops or as sources of natural enemies that move from one strip to another as in the case of alfalfa strips within cotton fields (Stern, 1979; Robinson, Young, and Morrison, 1972). In intercropping systems where crops are more closely intermingled, other mechanisms (i.e., repellency, masking, natural enemy enhancement, physical barriers) may affect insect pests. Clearly, much more work is needed to determine effective row spacings within crop mixtures that enhance pest suppression.

The manipulation of weed abundance and composition in intercrops can also have major implications on insect dynamics (Altieri, Schoonhoven, and Doll, 1977). When weed and crop species grow together, each plant species hosts an assemblage of herbivores and their natural enemies; thus, trophic interactions become very complex. As discussed in Chapter 4, many weeds offer important requisites for natural enemies such as alternate prey/hosts, pollen, or nectar as well as microhabitats that are not available in weed-free cropping systems (Van Emden, 1965b). Weed species that support rich natural enemy faunas include the perennial stinging nettle *(Urtica dioica),* Mexican tea *(Chenopodium ambrosioides),* camphorweed

(Heterotheca subaxillaris), and goldenrod *(Solidago altissima)*, as well as many weeds of the Compositae and Umbelliferae families. However, weed background may be important in other ways during the early stages of the intercrop when the pest arrives. Some evidence suggests that pests are deterred from ovipositing or remaining in crops with weedy background as opposed to bare soil. Also, many weeds may produce different chemical substances, thus confusing insects that localize their crops through chemical-feeding cues. Selective management of these weed species within intercrops may change the mortality of insect pests caused by natural enemies. The ecological basis for obtaining crop-weed mixtures that enhance insect biological suppression needs further development.

The final choice of cropping design must be dictated by the local nutritional needs and preferences, economic feasibility, and yield advantages of the mixture. Combinations of corn-legumes usually overyield corn monocultures; in other words, more area is needed under corn monocultures to produce the same yield as one hectare of polyculture (Vandermeer, 1981). This overyielding capability is expressed as a land equivalent ratio (LER). If higher than one, the ratio implies that the intercrop gives a better yield than the monoculture (Trenbath, 1976). The LER is defined as the relative land area required for sole crops to produce the same yields as intercropping. Using a simpler notation for competition studies, LER can be expressed as:

$$\text{LER} = \frac{Y_A + Y_B}{S_A + S_B}$$

Where Y_A and Y_B are the individual yields from crops in intercropping, and S_A and S_B are those for the same species as sole crops. Vandermeer (1981) discusses methods for interpreting data from intercropping and believes the LER to be a useful and practical indicator.

Chapter 6

Insect Ecology in Orchards Under Cover-Crop Management

Most biological control programs have been conducted in orchards and in protected environments because these are considered to be more stable and permanent ecosystems than annual-crop agroecosystems (Huffaker and Messenger, 1976). Several authors have claimed that insect populations are more stable in complex orchard communities because a diverse and more permanent habitat can maintain an adequate population of the pest and its enemies at critical times (van den Bosch and Telford, 1964). Orchards are semipermanent, relatively undisturbed systems, with no fallow; and crop rotation does not apply in the short term, so particular biological situations affecting insects occur in these systems.

For most entomologists, the relative permanency of orchards affords the opportunity of manipulating the components of an orchard habitat to the benefits of ecologically sound orchard management practices (Prokopy, 1994). One of these practices is the manipulation of ground-cover vegetation to enhance biological control of orchard arthropod pests.

In California, cover-crop management in orchards has been recommended because ground cultivation exacerbates soil erosion, reduces water penetration, and modifies the summer microclimate unfavorably (Finch and Sharp, 1976). Several legumes, such as lana vetch, clovers, and *Medicago* spp., and grasses such as brome, rye, and barley, have been recommended to be sown annually in orchards in the autumn or early spring, or at times farmers use cover crops that are self-regenerating and are thus sowed once. Cover crops are tilled or mowed yearly. Cover crops can add or retain soil nitrogen (N), facilitate the availability of their nutrients, produce organic matter, reduce soil compaction, improve water infiltration, and in some cases enhance moisture retention. In fact, cover crops can act as an "ecological turntable" influencing various agroecological processes simultaneously (Figure 6.1).

The manipulation of ground-cover vegetation in orchards and vineyards can significantly affect tree growth by altering nutrient availability, soil physics and moisture, and the prevalence of weeds, plant pathogens, and insect pests and associated natural enemies (Haynes, 1980). A great number of entomological studies conducted in these systems indicate that orchards with rich floral undergrowth exhibit a significantly lower incidence of insect pests than clean-cultivated orchards, mainly because of an increased abundance and efficiency of predators and parasitoids (Smith et al., 1996). Early in the twentieth century, Peterson (1926) observed that uncultivated orchards were attacked less severely by codling moth *(Cydia pomonella)* than were continuously cultivated orchards. Peppers and Driggers (1934) and Allen and Smith (1958) showed that the percentage of fruit moth larval parasitism was always greater in orchards with weeds than in clean-cultivated orchards.

In New Jersey peach orchards, control of the oriental fruit moth increased in the presence of ragweed (*Ambrosia* sp.), smartweed (*Poly-*

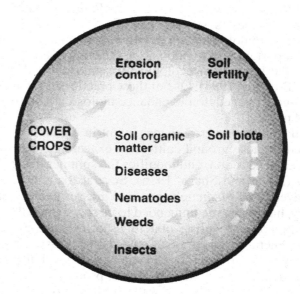

FIGURE 6.1. Multiple and interactive effects of cover crops on farming systems (after Michigan State University Bulletin E-2704, 2000).

gonum sp.), lamb's-quarter *(Chenopodium album)*, and goldenrod *(Solidago* sp.). These weeds provided alternate hosts for the parasite *Macrocentrus ancylivorus* (Bobb, 1939). Similarly, Leius (1967) found that the presence of wildflowers in apple orchards resulted in an eighteenfold increase in parasitism of tent caterpillar pupae over nonweedy orchards; parasitism of tent caterpillar eggs increased fourfold, and parasitism of codling moth larvae increased fivefold.

Considerable work was conducted by researchers in the Soviet Union on the role of nectar plants in increasing the effectiveness of biological control agents in orchards. Telenga (1958) reported that the parasitoid *Scolie dejeani* was attracted to its grub hosts when the honey plants *Phacelia* and *Eryngium* were sown. These same plants were shown to increase the abundance of the wasp *Aphelinus mali* for the control of apple aphids and to improve the activity of *Trichogramma* spp. wasps in apple orchards. Soviet researchers at the Tashkent Laboratory cited lack of adult food supply as a reason for the inability of *Aphytis proclia* to control its host, the San Jose scale *(Quadraspidiotus perniciosus)*. The effectiveness of the parasitoid improved as a result of planting a *Phacelia tanacetifolia* cover crop in the orchards. Three successive plantings of the *Phacelia* cover crop increased parasitization of scales from 5 percent in clean-cultivated plots to 75 percent in the *Phacelia* plots (Churnakova, 1960).

In the Solomon Islands, O'Connor (1950) recommended the use of a cover crop in coconut groves to improve the biological control of coreid pests by the ant *Oecophylla smaragdina subnitida.* In Ghana, coconut gave light shade to cocoa and supported, without apparent crop loss, high populations of *Oecophylla longinoda,* keeping the cocoa crop free from cocoa capsids (Leston, 1973).

Wood (1971) reported that in Malaysian oil palm *(Elaeis guineensis)* plantations, heavy ground cover, irrespective of type, reduced damage to young trees caused by the rhinoceros beetle *(Oryctes rhinoceros).* The mode of action is not certain, but it appears that the ground cover impedes flight of the adult beetles or restricts their movement on the ground. Economic control of this pest was possible by simply encouraging the growth of weeds between the trees.

Sluss (1967) reported another example of advantageous use of ground cover under trees. In California, the beetle *Hippodamia convergens* is the most important predator of walnut aphid *(Chromaphis juglandicola)* during the early season. This beetle moves from its overwintering area in the mountains to the walnut orchards in February and early March, when there are no leaves on the trees and therefore no aphids. However, some aphids are present in the ground cover under the trees and serve as a temporary food source for the predators which would otherwise move on or die of starvation. The ground cover under the trees should be chopped or disked in late April or early May to force the beetles onto the walnut trees. If it is chopped too early, however, the beetles will emigrate before the walnut aphids have appeared on the trees; if it is chopped too late, the large number of beetles will decimate the aphid population on the trees without ovipositing, resulting in fewer beetles later. Thus, timing of the chopping of the ground cover is critical to maintain ample beetle population for sufficient control of the aphids.

Fye (1983) proposed manipulation of orchard ground vegetation for building predator populations. In pear orchards of the Yakima Valley, he established various small grain and crucifer cover crops and found that several species of general predators were supported by aphids and *Lygus* bugs harbored by the cover crops.

In Michigan, ground plants can be allowed to grow up to the apple trees, since rainfall is sufficient for the trees not to suffer from competition for water by the grass. The phytophagous mites present on the cover constitute an early-season food source for the predatory mite *Amblyseius fallacis,* which later moves up into the trees and regulates the spider mites *Panonychus ulmi* and *Tetranychus urticae* (Croft, 1975).

Bugg and Waddington (1994) provide a list of understory weeds or "resident vegetation" that can become an asset when managed as cover crops as they harbor beneficial arthropods. Among the main species included are common knotweed, chickweed, toothpick ammi, sweet fennel, and sow thistle.

In China, Liang and Huang (1994) report that *Ageratum conyzoides* as well as other plants (*Erigeron annuus, Aster tataricus,* etc.) which encourage natural enemies, especially *Amblyseius* spp., of the citrus red mite *(Panonychus citri)* have been planted or conserved as ground cover in an area of 135,000 ha of citrus orchards with excellent results.

Also in China, Yan and colleagues (1997) developed a cover-crop system in apple orchards consisting of *Lagopsis supina* (Labiatae) instead of the traditional cover of Chinese rape *(Brassica campestris)* and/or alfalfa. *Lagopis supina* had a substantially greater effect in enhancing natural enemy populations than the other two plants.

The cover-crop work of McClure (1982), although not directed toward natural enemy enhancement, is proving useful for the manipulation of leafhopper pests in peaches. McClure's experiments demonstrated that ground cover significantly impacted the number of leafhoppers, *Scaphytopius acutus* (vectors of x-disease), colonizing peach trees. The great majority of adult leafhoppers occurred on trees with undergrowth of red clover and rosaceous weeds. Relatively few adults inhabited trees in plots with orchard grass, an unsuitable host. These data indicate that invasion of the orchard by leafhoppers can be discouraged by rendering the orchard floor free of naturally occurring wild host plants.

In California's Central Valley vineyards, variegated leafhopper population differences between cover and noncover plots were clear-cut for all three broods, but the reasons behind these differences were not so clear. Anecdotal reports from growers in the area suggest that weedy cover crops in early to midseason may result in smaller populations of leafhoppers. An increase in the abundance of generalist predators, especially spiders, may help reduce leafhopper populations in the weed-cover plots (Settle et al., 1986). In the same area, leaving a managed ground cover of johnsongrass or Sudan grass, a minor cultural practice modification in vineyards, resulted in a habitat modification which greatly enhanced the activity of predators against phytophagous mites such as the Willamette mite. When johnsongrass *(Sorghum halepense)* was allowed to grow in grape vineyards in California, there was a buildup of alternate prey mites, which supported populations of the predatory mite *Metaseiulus occidentalis,* which, in turn, restrained the Pacific mite, *Eotetranychus willamette,* to noneconomic numbers (Figure 6.2) (Flaherty, 1969).

Also in the San Joaquin Valley, the emergence of navel orangeworm adults *(Amyelois transitella)* was significantly higher in the complete, residual herbicide-treated almond orchards than in orchards with a vegetation cover. These results show that fewer navel orangeworms survive the winter on the ground if cover crops are present. The differences might be greater when nuts in cover crops are fully subjected to regular, early-spring mowing. Nuts in the residual herbicide treatments, which do not need mowing, would not be disturbed. Mowing, especially flail mowing, could reduce the navel orangeworm population further by physically destroying the overwintering nuts and larvae (Bugg and Waddington, 1994).

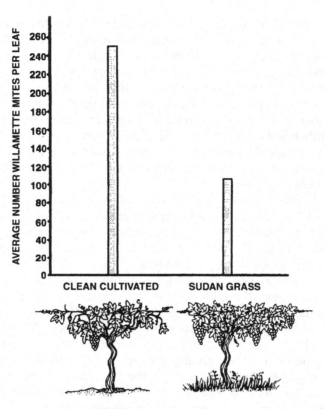

FIGURE 6.2. Effect of ground cover on Willamette mite populations in a California vineyard (after Flaherty, 1969).

SELECTING AND MANAGING COVER CROPS IN ORCHARDS

In cases where ground cover of any type is desirable, easily managed plants should be encouraged in preference to more aggressive species. In general, low-growing, nonclimbing legumes would be preferable to climbers or tall grasses, since climbers could cover the trees and tall grasses could limit movement between the trees. In addition, legumes such as vetch and fava beans fix more than 150 kg of N/ha and produce considerable amounts of biomass, an important input of organic matter into the orchard soil. If the ground cover will be regularly cut, it would be advantageous to use resident weeds that grow early in the spring and can regrow after repeated mowing or disking. These would probably be low-growing perennial grasses or perennial broad-leaved species, or a combination of the two. The advantage of weeds in'this situation is that they can often take much more abuse than cultivated species, and are thus easier to manage. There are, however, some cultivated legumes that might also be suitable for continual mowing.

The understory vegetation in an orchard need not be managed uniformly. Different zones may be treated differently; this is termed *strip management*, because the different treatments are usually applied linearly, and the different understory zones appear as bands or strips running through an orchard. Strip management of cover crops may entail (1) sowing cover crops of different floristic composition in different strips; (2) mowing strips at different times; (3) tilling strips at different times; or (4) combinations of these three processes. Sowing of different mixes leads to stands with differing statures and phenologies, thus affording diverse resources to pest and beneficial arthropods (Bugg and Waddington, 1994).

Wyss, Niggli, and Nentwig (1995) planted strips of weed mixtures to enhance populations of aphidophagous insects and spiders in Swiss apple orchards. During flowering of weeds more aphidophagous predators were observed on the apple trees within the strip-sown area than in the control area. The most abundant and permanent aphidophagous predators were spiders, predaceous Heteroptera, Coccinellidae, and Chrysopidae. Both species of aphids were significantly less abundant in the area with weed strips than in the control area during the vegetation period. Their results support strip management as a viable option to manage aphids.

Ideally, cover crops should be selected or managed so as to (1) not harbor important pests; (2) divert generalist pests; (3) confuse specialist pests visually or olfactory and thus reduce their colonization of orchard trees; (4) alter host-plant nutrition and thereby reduce pest success; (5) reduce dust and drought stress and thereby reduce spider mite outbreaks; (6) change the microclimate and thereby reduce pest success; and (7) increase natural enemy abundance or efficiency, thereby increasing biological control of arthropod pests (Bugg and Waddington, 1994).

CASE STUDY 1: APPLE ORCHARDS IN CALIFORNIA

During 1982 and 1983 a study was conducted on the effects of cover-crop manipulation on arthropod communities in three northern California apple orchards. The objectives were to (1) compare population levels and fruit damage by insects such as codling moths, aphids, and leafhoppers in orchards grown under clean cultivation or with cover crops, (2) find out whether undersowing cover crops in orchards would enhance populations of resident beneficial insects, and (3) evaluate the effects of cover-crop manipulation on tree growth and productivity.

The study comprised one disked orchard kept free of ground vegetation by one spring and one late summer disking. The other cover-cropped orchard was undersown in the fall with approximately twenty pounds of bell bean *(Vicia faba)* seeds per acre. By early June, the cover was mowed and the residues allowed to remain on the soil as straw mulch for the rest of the season.

The relative abundance of plant-feeding insects and associated natural enemies were monitored on five randomly selected trees per orchard, on cover crops, and on the orchard floor. In each orchard, the lower canopy of each tree was sampled for one minute with a D-Vac insect suction machine. A pitfall trap filled with 75 percent water and 25 percent antifreeze placed at the base of each tree captured ground-dwelling arthropods. Two Zoecon codling moth *(Cydia pomonella)* pheromone traps were placed in each orchard, indicating peak flights of male moths.

To assess codling moth damage at the end of each season, larval entries in 100 fruits collected from each of the five sample trees in each orchard were examined. All fruits from each tree were counted and weighed to determine yields per tree and percentage of total fruit damaged. Weekly evaluations of the proportion of infested twigs per tree indicated aphid and leafhopper levels.

Predation on tree foliage was assessed with twenty-five paper cards (3 inches × 4 inches), each containing fifty Mediterranean flour moth *(Anagasta kuehniella)* eggs per card, hung in five trees per orchard. Ground predation was estimated by 50 cardboard sheets (8.5 inches square), each with twenty glued potato tuberworm *(Phthorimaea operculella)* larvae, randomly placed on the floor of each orchard. The cards and sheets were removed after twenty-four hours and remaining eggs and larvae counted.

In 1982, substantially more male codling moths were caught in the disked orchard (a total of 275 caught on nine sampling dates) than in the cover-cropped orchard (164 moths). These differences did not occur during 1983. Densities of the rosy apple aphid *(Anuraphis roseus)* were slightly higher during May and June 1982 in the disked than in the cover-cropped orchard. In 1983, rosy apple aphids were detected only in the disked orchard, where they reached high numbers in early June. Leafhoppers (Homoptera: Cicadellidae) colonized the orchards late in the 1982 season, reaching substantially higher densities in the disked than in the cover-cropped orchard (Figure 6.3).

In both years, populations of natural enemies on the trees remained low, and no differences between orchards were apparent in seasonal abundance of common predators, such as Coccinellidae, Chrysopidae, and Cantharidae. Only spiders reached higher numbers on the trees with a cover crop early in the 1983 season. Despite these undetectable general differences in predator abundance, the number of *Anagasta* eggs removed from the trees was substantially higher in the cover-

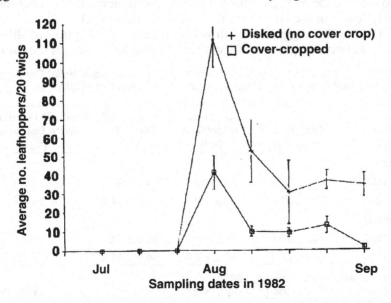

FIGURE 6.3. Leafhopper densities on apple trees with and without cover crop in California (after Altieri and Schmidt, 1986a).

cropped orchard than in the disked orchard, especially during July and August in both 1982 and 1983.

As expected, disking primarily affected ground predators. In both years, ants and spiders were caught in pitfalls more consistently in the cover-cropped orchard than in the disked orchard, and carabid ground beetles appeared to be more prevalent in the disked orchard, especially from July on. Predators (especially ants) appeared to be more effective in removing *Phtorimaea* larvae in the cover-cropped than in the disked orchard.

A variety of general predators and parasitic Hymenoptera (mainly larger Braconidae and Ichneumonidae) were present on the cover-crop vegetation. Most were supported by the high numbers of alternative prey (especially aphids) harbored by the cover-crop vegetation from early April through mid-June in both years. Leafhoppers were particularly prevalent on the cover crops in 1983.

In 1982, there were no apparent differences in fruit yields between the two orchards, but codling moth damage seemed slightly lower in the cover-cropped orchard. In 1983, however, cover-cropped trees produced considerably more fruit, although they were smaller, than did trees in the disked orchard. Codling moth incidence was substantially lower in the cover-cropped orchard (Table 6.1).

The results suggest the following tentative description of differences between apple systems with and without cover crops: Apple or-

TABLE 6.1. Apple Production and Codling Moth *(Cydia pomonella)* Damage in Organic Orchards with and Without Cover Crop in Northern California

Orchard and Year*	Total No. of Fruits/Tree	Total Fruit Weight/Tree (kg)	Fruit>2.5" Diameter (%)	Fruit with Codling Moth Damage (%)
Cover				
1982	241 ± 35.6**	29.1 ± 3.6	44.8 ± 9.8	68.0 ± 9.7
1983	334 ± 56.8	58.3 ± 9.7	87.6 ± 14.6	4.2 ± 0.7
Disked				
1982	260 ± 37.1	26.9 ± 4.5	39.0 ± 6.5	78.0 ± 9.7
1983	94 ± 11.7	15.5 ± 2.6	54.8 ± 11.0	38.9 ± 7.8

Source: Altieri and Schmidt, 1986b.
*Total rainfall during the growing season (April-October) was 341 mm in 1982 and 367 mm in 1983.
**Means ± SD

chards with cover crops generally had (1) lower infestation levels of aphids, leafhoppers, and codling moths, (2) more species and more individuals of soil-dwelling predaceous arthropods, and (3) higher removal rates of artificially placed prey. In contrast, disked systems were generally characterized by greater numbers of plant feeders on the trees and by relatively low population levels of natural enemies.

The cover crops generally harbored large numbers of prey, such as aphids and leafhoppers, which attracted varying numbers of predators. High numbers of predators on the cover crops, however, did not necessarily translate into higher numbers on the trees. Experiments to test whether the common practice of mowing the cover crop forces natural enemies to move up to the trees could be useful in designing management plans for encouraging efficiency of natural enemies. Although cover cropping significantly affected ground-predator populations, these studies could not determine how these changes affected pest species on the trees. The data also did not indicate how realistically predation on artificial baits related to reduction of apple pests, such as codling moths, aphids, and leafhoppers.

Depending on the orchard system, cover-crop complex, and associated arthropod species, it seems that manipulation of the ground cover can have a significant effect on the number of arthropods that inhabit the orchard by (1) directly affecting plant-feeding populations which discriminate between trees with and without cover underneath or (2) attracting and retaining soil- and foliage-inhabiting natural enemies by providing alternative food and habitats. Critical testing of these effects in a range of orchard systems may lead to improve biological control of certain orchard pests.

CASE STUDY 2: PECAN ORCHARDS IN GEORGIA

In southern Georgia, pecan trees are attacked by several aphid species, including yellow pecan aphid *(Monelliopsis pecanis),* black-margined aphid *(Monellia caryella),* and black pecan aphid *(Melanocallis caryaefoliae).* Through phloem feeding, these can reduce tree vigor and productivity. Moreover, late-season outbreaks may be prompted by insecticides used against other pests of pecan, and there is growing evidence of aphid resistance to available insecticides.

Winter legumes such as hairy vetch *(Vicia villosa)* and crimson clover *(Trifolium incarnatum)* are now being used in attempts to enhance early-season biological control of pecan aphids. This practice is an alternative to the prevalent understory-management approach in pecan that involves herbicide tree-row strips with intervening alleys of mown grass, to facilitate harvest by mechanical shakers and sweepers. These warm-season grasses employed (e.g., Bermuda grass *[Cynodon dactylon]* and centipede grass *[Eremochloa ophiuroides]*) harbor few beneficial insects and add little to soil nitrogen. Therefore, there is a need to develop a low-input, minimum-tillage scheme for year-round management of cover crops in the pecan agroecosystem. In order to meet this goal, Bugg and Dutcher (1989) conducted trials of several prospective warm-season cover crops that would serve as potential "insectary crops."

Among the various evaluated species Bugg and Dutcher (1989) found that *Sesbania exaltata* was the best source of cowpea aphid and the best reservoir for various hover flies and coccinellid beetles. Cowpea aphids on *Indigofera hirsuta* attracted various aphidophagous predators in November after pecan leaves were shed. They further found that during the summer, the flowers of buckwheat *(Fagopyrum esculentum)* attracted many entomophagous wasps. The understory of sesbania cover crops supported high densities of banded-winged whitefly and cowpea aphid, which were colonized by cocinellids *(Olla v-nigrum* and *Hippodamia convergens).*

In another study, Bugg and Waddington (1994) found that in mature pecan orchards under minimal or commercial management, cool-season understory cover crops of hairy vetch and rye sustained significantly higher densities of aphidophagous lady beetles than did unmown resident vegetation or mown grasses and weeds. In cover-cropped understories, mean densities of aphidophagous coccinellids were nearly six times greater than in unmown resident vegetation and approximately eighty-seven times greater than in mown grasses and weeds.

Dutcher (1998) considers cover crops key components of pecan integrated pest management (IPM) combined with reduced frequency of pesticide sprays, planting legumes as intercrops in the orchard to produce alternate prey aphids for aphidophaga, and partitioning of the foraging behavior of the red imported fire ant with trunk sprays of insecticide that prevent ants from reaching aphids and mealybugs in the tree, yet allowing ants to remain on the orchard floor as predators of pecan weevil larvae. Years of experimentation show that inter-cropping with sesbania alone or the combination of intercropping with hairy indigo and ant exclusion reduce pecan aphid populations in Georgia.

CASE STUDY 3: SUMMER COVER CROPS IN VINEYARDS

In California, some researchers have tested planting cover crops as a habitat management tactic in vineyards to enhance natural enemies, including spiders (Costello and Daane, 1998). Reductions in mite (Flaherty, 1969) and grape leafhopper (Daane et al., 1998) popula-tions have been observed, but such biological suppression has not been sufficient from an economic point of view (Daane and Costello, 1998). Perhaps the problem lies in the fact that most of these studies were conducted in vineyards with winter cover crops and/or with weedy resident vegetation which dried early in the season or which was mowed or plowed under at the beginning of the growing season. Therefore, in early summer these vineyards are virtual monocultures without floral diversity. For this reason Nicholls, Parrella, and Altieri (2000) tested the idea to maintain a green cover during the entire growing season in order to provide habitat and an alternate food for natural enemies. They sowed summer cover crops (buckwheat and sunflower) that bloom early and throughout the season, thus provid-ing a highly consistent, abundant, and well-dispersed alternative food source, as well as microhabitats, for a diverse community of natural enemies.

This study was conducted in two identical adjacent Chardonnay organic vineyard blocks from April to September in 1996 and 1997. Vineyards were located in Hopland, 200 km north of San Francisco, California. One block was kept free of ground vegetation by one spring and one late-summer disking (monoculture vineyard). In April, the other block (cover-cropped vineyard) was undersown in every al-ternate row with a 30/70 mixture of sunflower and buckwheat. Buck-wheat flowered from late May to July, and as the buckwheat senesced, sunflower bloomed from July to the end of the season.

These researchers found that maintenance of floral diversity throughout the growing season in vineyards in the form of summer cover crops had a substantial impact on the abundance of western grape leafhoppers, *Erythroneura elegantula* (Homoptera: Cicadellidae), western flower thrips, *Frankliniella occidentalis* (Thysanoptera: Thripidae), and associated natural enemies.

During two consecutive years, vineyard systems with flowering cover crops were characterized by lower densities of adult and nymph leafhoppers (Figure 6.4) and thrips, and larger populations and more species of general predators, including spiders. Although *Anagrus epos* (Hymenoptera: Mymaridae), the most important leafhopper parasitoid, achieved high numbers and inflicted noticeable mortality of grape leafhopper eggs, no differences in egg parasitism rates were observed between cover-cropped and monoculture systems. Mowing of cover crops forced movement of *Anagrus* and predators to adjacent vines resulting in the lowering of leafhopper densities in such vines (Figure 6.5).

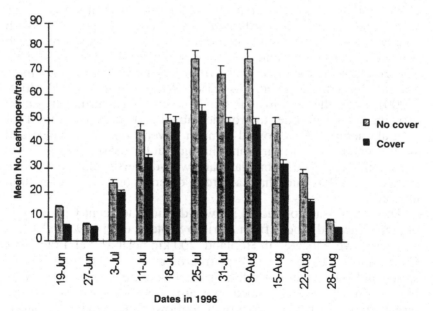

FIGURE 6.4. Densities of adult leafhoppers *E. elegantula* in cover-cropped and monoculture vineyards in Hopland, California, during the 1996 growing season. Mean densities (number of adults per yellow sticky trap) and standard errors of two replicate means are indicated. In some cases error bars were too small to appear in the figure (after Nicholls, Parrella, and Altieri, 2000).

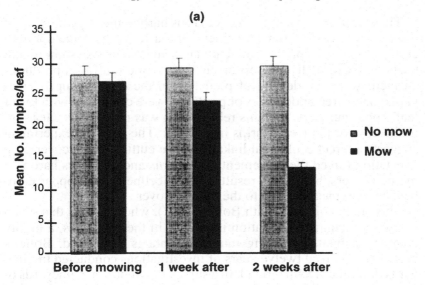

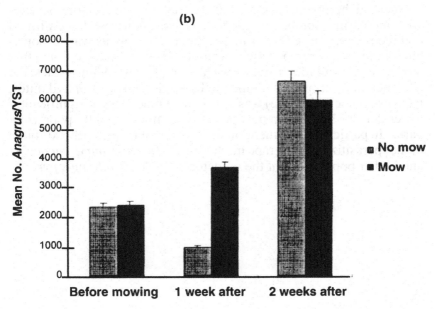

FIGURE 6.5. Effects of cover-crop mowing in vineyards on densities of leaf-hopper nymphs and *Anagrus epos* during the 1997 growing season in Hopland, California (after Nicholls, Parrella, and Altieri, 2000).

These studies showed that cover crops harbored a large number of *Orius,* coccinellid, thomisid spiders, and a few other predator species. Comparisons of predator abundance in both blocks showed that the presence of such predators on buckwheat and sunflower produced an increase in the density of predators in the cover-cropped vineyards. Such greater densities of predators were correlated with lower leafhopper numbers, and this relationship was much more clear-cut in the case of the *Orius*-thrips interaction. The mowing experiment suggests a direct ecological linkage, as the cutting of the cover-crop vegetation forced the movement of *Anagrus* and predators harbored by the flowers, both years resulting in a decline of leafhopper numbers on the vines adjacent to the mowed cover crops.

This study coincides with Boller (1992), who reported that when understory summer vegetation is present in the vineyards, a highly complex habitat containing multiple strata is developed. Boller's group summarized twelve years of investigations conducted in vineyards of northern Switzerland. In a survey of twenty-one vineyards of varying flower richness these authors found a considerable increase in "neutral herbivores" and of beneficial entomophagous species with increasing numbers of plant species. Their results confirmed that the proportion of flowering perennial dicot plants was responsible for the increase in beneficial arthropod taxa. They also found that in vineyards exhibiting flora with a large number of plant species, the tendency in most species (grape moths, spider mites, eirophyd mites, thrips, and noctuid larvae) was to fluctuate much less at significantly lower density levels than pest populations in botanically poor vineyards. In particular they mention that flower-rich vineyards exhibited higher parasitization of grape moth eggs by *Trichogramma cacoeciae* and higher populations of the predatory mite *Typhlodromus pyri.*

Chapter 7

The Influence of Adjacent Habitats on Insect Populations in Crop Fields

The biogeographic region rather than the single homogeneous field may often be the appropriate unit for pest-management research (Levins and Wilson, 1979). According to Rabb (1978), an agroecosystem should be conceived as an area large enough to include those uncultivated landscapes that influence crops through inter-community interchanges of organisms, materials, and energy. Our single-commodity approach in organizing research on pests often allows us to ignore associations with other crops, host plants, and adjacent plant communities that are of critical importance in the life systems of pests.

The vegetational component of agroecosystems can be viewed as a mosaic of annual and perennial crop fields, forest patches, pasturelands, fallow fields, orchards, swamps, old fields, and tree plantations. The agricultural landscape consists of (1) the agricultural field (consisting usually of a single crop and any weeds present, but sometimes including additional crops or a cover); (2) native and/or weedy vegetation that may be present on its borders; (3) the surrounding agricultural fields; and (4) the vegetation occurring in native or uncultivated habitats in the surrounding area. The composition of the agricultural landscape determines the presence of overwintering sites and the ability of an insect to locate appropriate habitats and food resources over the course of its lifetime (Perrin, 1980). Although it is useful to consider agroecosystems as a crop "island" subject to colonization from several sources, a regional perspective is necessary for predicting the movement patterns of pests and natural enemies across agricultural landscapes. Rabb (1978) argues that an insect population's performance and survival is related to the large-unit ecosystem heterogeneity, especially when the risks assumed by the population in moving between different sites are considered. Root (1975) discusses "compound ecosystems" and the variable responses of herbivores to the dispersion and size of resource patches within habitats. Many predators tend to feed on several different species of prey and to distribute themselves on vegetation in response to the availability of prey rather than the plant species. On a broad scale, just as the heterogeneity of distributions of plant populations can influence the effects of herbivores on plants, so the heterogeneity of herbivore distribution patterns can influence the effects of predators and/or parasites on the herbivores.

Several studies suggest that the vegetational settings associated with particular crop fields influence the kind, abundance, and time of arrival of herbivores and their natural enemies (Price, 1976). Large populations of certain pests, especially polyphagous, univoltine species, migrate en masse from alternate hosts in the vicinity to newly established (and presumably, vulnerable) crop monocultures (Duelli et al., 1990). The scale and intensity of this phenomenon depends, of course, upon the vagility of the insects involved (Andow, 1983b).

Insects are mobile species; the scale of their home range differs according to their method of locomotion (walking or flying) and dispersal. Highly mobile organisms may use different fields or uncultivated areas during their life. Studies of the ecology of such organisms must consider processes not only on individual sites but also at the regional or landscape level. The diversity of the farmland mosaic, defined by the variety of crops and wild plants, and their spatial arrangement, for example, the size of fields and the heterogeneity of their spatial distribution, play key roles in determining the abundance, diversity, and dispersion of insect species (Baudry, 1984). The movement of individual insects will respond to a wide range of landscape factors including the scale of the habitat (plant, field, and landscape level), habitat permeability, patch size and shape, and degree of isolation (Figure 7.1). Landscape structure influences microclimate and crop growth, as well as other factors that affect the movement patterns of insects.

Studies in North Carolina have shown that field size and geographical · distribution of fields can affect the movement and location of Mexican bean beetle, *Epilachna varivestis,* infestation (Stinner et al., 1983). Since this insect overwinters along the edges of woods and reproduces much faster on garden beans than on soybean, the beetles exhibit a seasonal movement from woody edges to beans to soybean and back to woods.

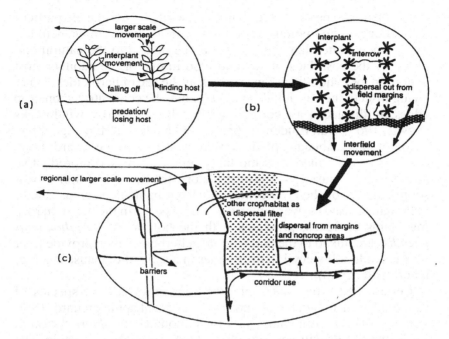

FIGURE 7.1. Insect movement in response to landscape structure on arable farms shown at three scales: (a) the scale of the individual plants; (b) the field scale; and (c) the landscape scale. The size, shape, and spatial pattern of patches will be important across all of these scales (after Fry, 1995).

Significant movement against the direction of prevailing winds is possible by strong fliers or simply by alighting until the winds change from the prevailing direction. Johnson, Turpin, and Bergman (1984) describe an example of southerly movement by *Leptinotarsa decemlineata,* the Colorado potato beetle, in the same area that *Empoasca fabae* moves from spring breeding grounds in the lower Mississippi basin to Wisconsin crop fields with the prevailing air currents.

CROP EDGES AND INSECT PESTS

Several species of weeds present in areas adjacent to crops may serve as alternative hosts for crop pests (Van Emden, 1965b). Most of these insect pests tend to feed on wild plants botanically related to the crops. Over 200 pests of cereals utilized wild grasses that are particularly abundant and ubiquitous in arable areas. Insect movement between uncultivated land and crops can be related to the natural dispersal of the pest, lack of suitable food in one of the habitats, host alteration, or major disturbances, such as the use of herbicides. Small flying insects such as aphids and thrips may be concentrated on crop edges by displaced air currents attributable to nearby windbreaks (Lewis, 1965). Van Emden (1965a) found a heavy initial edge infestation of alate cabbage aphids caused by shelter to windward; however, because of increased mortality and a decreased reproductive rate probably caused by physical factors, the density of aphids was soon half that of the crop center. Adjacent habitats, such as shelter belts, can be used as overwintering sites by pests. In the northern rolling plains of Texas, problems with the boll weevil *(Anthonomus grandis)* are linked to the planting of shelterbelts which provide litter for the adult boll weevil to overwinter in a state of diapause (Slosser and Boring, 1980).

Certain hedgerow plants provide sources of several species of pests and predators that may move into adjacent apple orchards (Solomon, 1981). In England, the winter moth *Operophtera brumata* feeds on wild *Prunus* spp., beech, and oak, in addition to apple. The larvae can disperse by drifting in the wind, so hedgerow and woodland trees can be significant local sources of this pest. The underlying herb layer of a woody border may include plants attractive to crop pests. Wainhouse and Coaker (1981) found that the distribution of the perennial stinging nettle, *Urtica dioica,* explained the abundance of carrot fly, *Psila rosae,* in noncrop borders and have suggested a strategy to simplify field boundaries in order to minimize fly infestations.

On the basis of his survey of crop-field borders, Dambach (1948) concluded that the more nearly border vegetation is related, botanically, to the adjacent crop plants, the greater is the danger of its serving as a potential source of infestation by injurious insects. Thus, less crop-pest risk is involved in the use of woody border vegetation in areas where the predominant crops are grain, vegetables, and forage plants.

In temperate regions, an increasingly common alternative to hedgerows around orchard margins is the single-species planted windbreak. Poplar (*Populus* spp.), willow (*Salix* spp.), and some conifers are used, but the most widespread is alder (*Alnus* spp.). None of these windbreak trees provides an important source of phytophagous insects or mites that feed on apple, so they pose no threat to orchard-pest management (Solomon, 1981).

FIELD BOUNDARIES AND NATURAL ENEMIES

There is clear evidence that plants outside or around the cultivated field provide important resources to increase the abundance and impact of natural enemies. Habitats associated with agricultural fields may provide resources for beneficial arthropods that are unavailable in the crop habitat, such as alternate hosts or prey, food and water resources, shelter, favorable microclimates, overwintering sites, mates, and refuge from pesticides (Dennis and Fry, 1992).

Generally, hedges support a richer insect community than adjacent crop fields (Lewis, 1965), and the presence of certain hedges can enrich the insect population nearby for approximately a distance to leeward of three to ten times its height and to windward of equal to or twice its height.

Since the studies of Dambach (1948), it is known that the shelter provided by edge vegetation is important in encouraging natural enemies. There are still, however, many questions that need further research (Wratten, 1987; Kajak and Lukasiewicz, 1994):

1. To what extent do beneficial insects depend on hedges, ditch banks, old fields, and forests for their continued existence in agricultural areas, particularly during winter?
2. Do these borders and other abrupt transitions between one ecosystem and another (ecotones) influence the species diversity and abundance of entomophagous insects in adjacent crop fields?
3. Which attributes of the boundary are important for the natural enemies?
4. Can existing natural refuges within boundaries be improved or can new refuges be created?

Several researchers have shown that vegetation in adjacent areas can provide the alternative food and habitat essential to perpetuate certain natural enemies of pests in crop fields. Many beneficial insects find overwintering quarters in the litter developed in shrub and osage *(Maclura pomifera)* borders of crop fields (Dambach, 1948). Van Emden (1965a) and Pollard (1968) showed that the proportion of predaceous insects increased with reduced hedgerow management.

In his study, Pollard (1968) divided a 4 m high hawthorn hedge adjacent to a cereal field into six 30 m lengths and removed the bottom flora from three 30 m sections with a paraquat-diquat mixture, a total weed killer. Both sides of the hedge were treated, and the treatment continued for three seasons. This removal of shelter (as well as, of course, herbaceous plants harboring alternative prey) significantly reduced the predatory fauna in the bottom 1.5 m of the hedgerow. *Anthocoris nemorum* was one predator affected, and so were spiders and the carabid beetles *Bembidion guttula* and *Agonum dorsale*. These predators were particularly likely to colonize the adjacent fields.

In more detailed studies with *Agonum dorsale,* Pollard (1968) was able to show that this predator overwintered in the hedge bottom. By dissecting females and examining their ovaries, he showed that *A. dorsale* captured prey 54 m into pea and wheat crops, as it probably invaded the fields from the hedge.

Several other studies indicate that the abundance and diversity of entomophagous insects within a field are related closely to the nature of the surrounding vegetation. In northern Florida, predator density and diversity were greater in maize plots surrounded by annually burned pinelands and complex weedy fields than in those plots surrounded by sorghum and soybean fields (Altieri and Whitcomb, 1980). Neighboring vegetation can also determine the rates of colonization and population gradients of natural enemies within a particular crop field (Altieri and Todd, 1981). A study of the dispersal of carabid and staphylinid adults into cereal fields from field boundaries (Coombes and Sotherton, 1986) showed that beetles could be recovered up to 200 m into the fields and that two patterns of dispersal could be distinguished. One pattern typical of the carabids *A. dorsale* and *Tachyporus hypnorum* showed decreasing numbers, with progressively later peaks along a transect from the boundary to the center of the field (Figure 7.2).

Similarly, in southern Georgia, predators were more abundant in the edges of soybean fields adjacent to pea fields and weedy tracts than in soybean edges adjacent to vegetation-free fields (Altieri and Todd, 1981). In Georgia, predator numbers in soybean rows declined sharply the farther the rows were from a weedy ditch bank and a for-

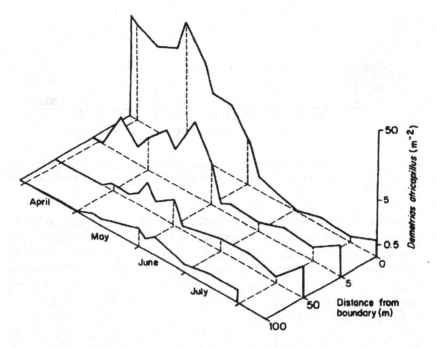

FIGURE 7.2. Density of the ground beetle *(Demetrias atricapillus)* at different distances from a field boundary (after Wratten, 1987).

est edge adjacent to the field (Figure 7.3). Van Emden (1965a) found that syrphid predators of the cabbage aphid, *Brevicoryne brassicae,* were distributed in crop edges near flowering weeds; predation kept pest densities at the crop borders to below half the level found at the center of the crop area.

In Hawaii the presence of nectar-source plants in sugar cane field margins allowed population levels to rise and increased the efficiency of the sugar cane weevil parasite *Lixophaga sphenophori* (Topham and Beardsley, 1975). The authors suggest that the effective range of the parasite within cane fields is limited to about 45 to 60 inches from nectar sources present in the field margins. Continuous herbicidal elimination of field-margin nectar-source plants had a detrimental effect on populations of *Lixophaga* and, therefore, led to a decrease in the efficiency of the parasite as a biocontrol agent of the weevil (Table 7.1).

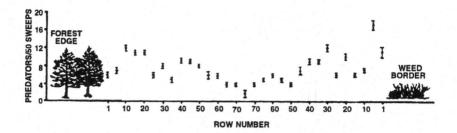

FIGURE 7.3. Abundance gradient of insect predators across a soybean field adjacent to two different types of plant communities in Georgia (after Altieri and Todd, 1981).

Maier (1981) observed higher parasitization rates of apple maggot *(Rhagoletis pomonella)* by braconids in northern Connecticut apple and hawthorn orchards where plants such as blueberry *(Vaccinium* spp.), dogwood *(Cornus* spp.), and winterberry *(Ilex cillata)* commonly grew nearby. These plants support populations of several frugivorous tephritids that serve as alternate hosts to the braconids.

In Norway's apple orchards, the numbers of the key pest *Argyresthia conjugella* is largely dependent on the amount of available food, i.e., the number of berries of the wild shrub *Sorbus aucuparia* that develop each year. Since only one larva develops in a single berry, the number of *Argyresthia* can never be higher than the total number of berries. Thus, in years when *Sorbus* has no berries in a certain area, no *Argyresthia* larvae are produced, and consequently there will be no parasites (the braconid *Microgaster politus*) in the area. Entomologists have suggested plantings of *Sorbus* which produce an abundant and regular crop every year. *Argyresthia* always finds enough food to maintain its population at a reasonably high level. Under such conditions, *Microgaster* and other natural enemies will also operate and reproduce sufficiently every year to regulate their host below the level where *Argyresthia* is forced to emigrate. Hence, the apple avoids infestation (Edland, 1995).

In a two-year study, Landis and Haas (1992) found higher parasitism of *O. nubilalis* larvae by the parasitoid *Eriborus terebrans* (Hymenoptera: Ichneumonidae) in edges of cornfields than in field interiors; and in the second year, they observed significantly higher parasitism by *E. terebrans* in edges of corn plantings adjacent to

TABLE 7.1. Parasitization of *R. obscurus* Grubs by *Lixophaga sphenophori* Before and After Application of Herbicide to Field Margin

Time of Test	Distance Infield from Margin	Percent of recovered grubs parasitized	
		Herbicide-Treated Field	Untreated Control Field
Before herbicide application	margin	95.0	100.0
	50 ft.	100.0	85.6
	100 ft.	100.0	87.5
	150 ft.	94.4	100.0
	200 ft.	26.4	100.0
	TOTAL	80.7	95.0
Immediately after herbicide application	margin	79.0	89.0
	50 ft.	86.5	71.4
	100 ft.	83.4	89.0
	150 ft.	92.8	70.0
	200 ft.	23.6	100.0
	TOTAL	76.5	83.6
35 days after herbicide application	margin	23.6	60.0
	50 ft.	5.3	71.4
	100 ft.	0.0	62.5
	150 ft.	16.7	50.0
	200 ft.	5.6	73.0
	TOTAL	10.0	64.9

Source: after Topham and Beardsley, 1975.

wooded areas than in those areas near nonwooded areas or in field interiors. In Germany, parasitism of rape pollen beetle was about 50 percent at the edge of all fields. Toward the center of the field it dropped significantly to 20 percent (Thies and Tscharntke, 1999). In their studies in Illinois, Mayse and Price (1978) found that the mean number of both herbivore and predator/parasitoid species per habitat space in soybean fields was higher at the edge than in the center of the field. The presence of relatively complex vegetation in the crop borders was an important factor in such trends.

Carabids have been studied by many researchers in central Europe who concluded that important ground beetle predators of crop pests exploit the shelter of hedges during field cultivation and during winter. Pollard (1968) argues that in many sparsely forested countries much of the carabid crop fauna is of woodland origin and that many species are now dependent to a large extent on hedges for their continued existence in agricultural areas. *Agonum dorsale* is a carabid that exhibits seasonal migration between field and edge. Sotherton (1984) observed that different predatory Carabidae and Staphylinidae preferred different field-boundary types but that hedge banks or shelterbelts were more attractive to most polyphagous predators than grass banks or grass strips. Wallin (1985), also investigating the spatial and temporal distribution of Carabidae in cereal fields and adjacent habitats, suggested that certain carabid species seemed to prefer field edges at different times of their lives. The field edges serve as important shelter habitats at certain times of the season and function importantly as overwintering sites for various carabid species. More recently, Varchola and Dunn (1999) convincingly showed that both simple and complex roadside vegetation bordering cornfields was important to the carabid communities, especially before canopy closure in cornfields. Such habitats apparently provide carabids with necessary resources and functions, especially overwintering and breeding sites that are unavailable in relatively bare crop fields. In the Rhineland, Thiele (1977) observed that following the grain harvest on August 5, the catches of *A. dorsale* on the edges rose rapidly during August from the initial minimum values in July. Based on several studies of ground beetle movements and beetle habitat requirements, conceptual models of beetle movement between boundaries and adjacent crop fields have been proposed (Figure 7.4). Chiverton and Sotherton (1991) studied the effects of excluding herbicides from cereal-crop field edges; they were able to demonstrate that these edges had greater abundance of nontarget arthropods and thus supplied predators with ample prey. All these earlier studies of carabid habitat preferences and relative abundance led to research directed at conserving and enhancing carabid populations in and around annual crop fields. The same applies to other arthropods, and LeSar and Unzicker (1978) have proposed the establishment of grass or legume strips along field margins to enhance the colonization of soybean fields by spiders.

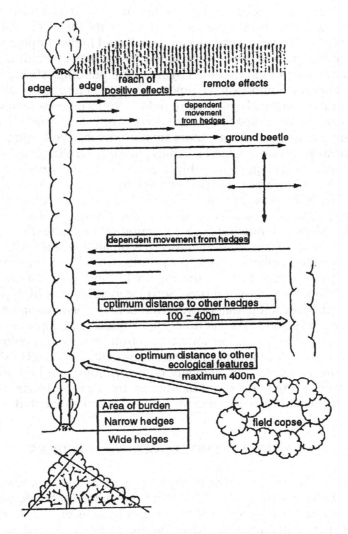

FIGURE 7.4. A conceptual model of carabid beetle movement between fields and hedges (after Thiele, 1977).

The proximity of forest edges and hedgerows that serve as hibernation sites has a fundamental effect on the occurrence of coccinellids in agricultural areas. In Czechoslovakia, an apple orchard surrounded by deciduous forests had a tenfold higher abundance of *Coccinella quinpuepunctata,* because the neighboring forest provided a dormancy site for the beetle, which hibernates in the litter of forest edges (Hodek, 1973). In England, researchers found higher populations of coccinellids in bean plots surrounded by patches of nettles, as opposed to bean plots surrounded by trees and buildings (Burn, 1987).

Several other authors have reported that the presence of alternate hosts or prey on weeds growing in field margins increases parasitism and/or predation of specific pests within crops. Since the life cycles of many parasitoids and predators are not synchronized with those of their hosts/prey, some natural enemies must rely on alternative sources to maintain establishment within a community. This may be especially important when the pest species have become scarce in the field. The widespread planting of alder (*Alnus* spp.) in southeast England has established a considerable reservoir of the predacious mirid *Blepharidopterus angulatus,* a regulator of the mite *Panonychus ulmi.* On these trees, *B. angulatus* feeds on aphids and leafhoppers; and when the numbers of these prey decline in August, they move on to nearby orchards, thus controlling spider mite populations (Solomon, 1981). Flowering willows, *Salix caprea,* support high populations of the predacious anthocorids *Anthocoris nemorum* and *A. nemoralis* in early April. At this time, apple aphids and apple psylla are just beginning to hatch, and anthocorids from willow may colonize in response to high numbers of these species. Aveling (1981) showed that *A. nemorum* was an abundant predator, migrating into hop gardens from spring populations in adjacent trees and hedgerows, especially from nettles, *Urtica dioica,* infested with nettle aphid.

DESIGNING AND MANAGING BORDERS

Klinger (1987) provided margin strips of *Sinapsis arvensis* and *Phacelia tanacetifolia* and found that these led to higher densities of polyphagous predators in the strips and in adjacent fields than in wheat plots without strips. Also, syrphid adults occurred at higher densities in the strips than in the field, presumably because the flies foraged on *P. tanacetifolia* and *S. arvensis.* The impact of different predatory groups on aphid populations was not quantified in this work, although there was a trend toward lower aphid densities in the field with adjoining strips. Sengonca and Frings (1988) found syrphid adults to be more abundant in sugar beet plots with *P. tanaceti-*

folia margin strips than in sugar beet monocultures. Borders of *Phacelia* have also been explored in cabbage where syrphid numbers increased, and aphid populations declined. Speight (1983) cites work reporting that strips of dill and coriander in eggplant fields led to enhanced predator numbers (*Coleomegilla maculata* and *Chrysoperla carnea*), increased consumption rates of Colorado potato beetle *(Leptinotarsa decemlineata)* egg masses, and decreased larval survivorship.

In many cases, weeds and other natural vegetation around crop fields harbor alternate hosts/prey for natural enemies, providing seasonal resources to bridge the gaps in the life cycles of entomophagous insects and crop pests. A classic case is that of the egg parasitoid wasp *Anagrus epos* whose effectiveness in regulating the grape leafhopper, *Erythroneura elegantula,* was increased greatly in vineyards near areas invaded by wild blackberry *(Rubus* sp.). This plant supports an alternative-host leafhopper *(Dikrella cruentata),* which · breeds in its leaves in winter (Doutt and Nakata, 1973). Recent studies have shown that prune trees planted next to vineyards also allow early-season buildup of *Anagrus epos.* After surviving the winter on an alternate host, the prune leafhopper, *Anagrus* wasps move into the vineyard in the spring, providing grape leafhopper control up to a month earlier than in vineyards not near prune tree refuges (Murphy, Rosenheim, and Granett, 1996). Murphy and colleagues (1998) completed a rigorous evaluation of the effectiveness of French prune trees in increasing control of the grape leafhopper. Results from this study indicate that there is a consistent and significant pattern of higher parasitism in grape vineyards with adjacent prune tree refuges than in vineyards lacking refuges. Researchers now recommend that trees should always be planted upwind from the vineyard but otherwise can be managed as a typical commercial prune orchard; they also suggest planting as many trees as is economically feasible, since the more trees there are, the more productive the refuge is likely to be.

By monitoring rubidium-labeled *Anagrus,* Corbett and Rosenheim (1996) found that *Anagrus* colonizing the study vineyards from external sources consistently exhibited a distinct spatial pattern: low abundance in the first vine row downwind of French prune trees; a large increase at the third vine row downwind; and a gradual decline from this peak with increasing distance from the refuge. It is likely that a windbreak effect is operating in this system: *Anagrus* emerging from overwintering habitats external to the vineyard-French prune system are colonizing at a higher-than-average rate immediately downwind of the refuge as a result of the turbulence generated by French prune trees. French prune tree refuges are thus having two impacts on early-season abundance of *Anagrus:* (1) directly contributing *Anagrus* that have overwintered in the refuge and (2) increasing the colonization rate by *Anagrus* having overwintered in habitats external to the French prune-vineyard system (Figure 7.5). The amount of additional colonization generated by a windbreak effect of refuges is dependent on the proximity and size of external overwintering habitats, because *Anagrus* must be dispersing in large numbers in the wind stream for a windbreak effect to cause increased colonization. Thus, refuges that are close to riparian habitats would generate high colonization, whereas refuges that are many kilometers away might generate no noticeable windbreak-induced colonization.

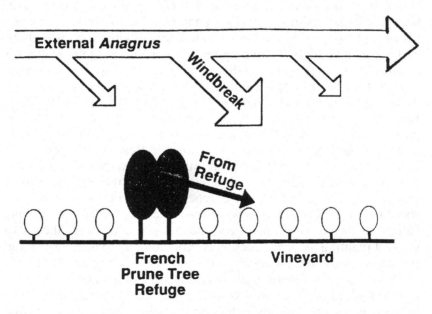

FIGURE 7.5. Hypothesized sources of *Anagrus* colonizing vineyards early in the season. *Anagrus* colonize vineyards from adjacent French prune tree refuges. *Anagrus* also colonize from external overwintering sites. The windbreak effect generated by prune trees causes increased colonization by external *Anagrus* immediately downwind of refuges (after Corbett and Rosenheim, 1996).

At a more regional level, one of the unique cases exploring the relationship between landscape, vegetation diversity, and insect pests comes from a twenty-year experiment conducted near Waco, Texas, from 1929 to 1949 by the Bureau of Entomology and Plant Quarantine in cooperation with the Soil Conservation Service. Entomologists measured the effects of new farming and soil conservation methods on populations of beneficial and pest insects in cotton (De Loach, 1970). About 600 acres of upland farmland was divided into adjacent areas of 300 acres each, designated Y and W (Figure 7.6). In both, the old farming practices were continued for four years, through 1942, while pretreatment counts were made. New conservation methods then were begun in Y, while the old practices were continued in W, where cotton occupied the largest acreage, followed closely by corn. There was also a substantial acreage of oats and pas-

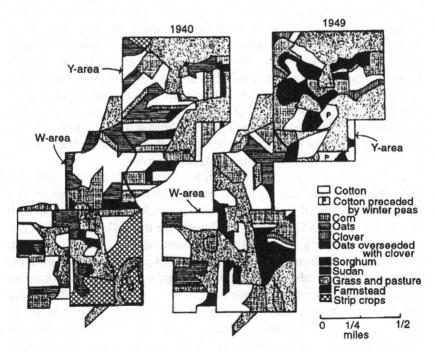

FIGURE 7.6. Diversification of agricultural landscapes in Texas for soil conservation purposes that led to improved control of insect pests of cotton (after De Loach, 1970).

ture and a little sorghum; these crops occupied nearly 100 percent of the land area. By the new practices, several acres of clover were planted alone or overseeded in the oats, some grass areas were added, and the land was terraced. Insecticides were not used in any fields in the two areas during the experiment, so that the effect of cultural methods alone could be measured. The new conservation practices resulted in a reduction in numbers of cotton pest insects and a reduction in the percentage of damaged squares and bolls. No attempt was made at elucidating the mechanisms explaining such reductions, although it is assumed that natural enemies were highly favored by the new landscape designs, thus resulting in enhanced pest mortality.

More recent research conducted in the mid-1990s suggests that within agricultural fields the diversity of parasitoids and the intensity of parasitism is generally greater at field edges where crops are adjacent to later-successional plant communities than in field interiors or along field edges with crop-crop or crop-early successional interfaces. Also, on a larger spatial scale, the diversity of parasitoids and the intensity of parasitism should be greater in agricultural landscapes embedded in a matrix of later-successional plant communities (old fields, hedgerows, woodlots) than in simple agricultural landscapes composed primarily of field crops. Therefore, on both the small within-field and the large between-field scales, a highly diverse landscape structure may provide the greatest potential for the biological suppression of pests by their natural enemies.

Marino and Landis (1996) compared maize fields of small size embedded in a landscape of abundant hedgerows and woodlots to a simple landscape of large-size fields embedded in a landscape with few hedgerows and woodlots to determine the influence of overall landscape diversity on parasitoid communities of the armyworm *(Pseudaletia unipunctata)*. They found that parasitism was greatest in the complex landscape. *Meteorus* wasps were the most abundant parasitoids attacking armyworm, and the presence of alternative hosts in the complex landscape explained its increased abundance. In a study in northern Germany, Thies and Tscharntke (1999) found that structural simplicity in agricultural landscapes was correlated with large amounts of plant damage caused by the rape pollen beetle *(Meligethes aeneus)* and small amounts of larval mortality caused by three ichneumonid parasitoids.

These studies give credence to emerging approaches that suggest the importance of the landscape as a level of organization of processes such as dispersion of plants, arthropod movement, and nutrient flow (Paoletti, Stinner, and Lorenzoni, 1989). Since agriculture is a major force shaping landscape structure and dynamics, it is useful to examine the relationships between arthropods and vegetation patterns at the landscape ecological level, especially in regions dominated by large-scale monocultures, which represent highly fragmented landscapes. Much concern has been expressed regarding the effects of these fragmented landscapes on the survival of a variety of beneficial entomofauna. As habitats become more fragmented, a variety of species that require relatively large areas of suitable habitat have a more difficult time surviving in the increasingly smaller fragments; populations may become extinct in these fragments, and they may never return. Reversing these effects through agricultural diversification is a key challenge.

CASE STUDY 1: EXCHANGE OF ARTHROPODS AT THE INTERFACE OF APPLE ORCHARDS AND ADJACENT WOODLANDS

In northern California, apple orchards are distributed within a matrix of natural vegetation that provides abundant opportunities to study arthropod colonization and interhabitat exchanges of arthropods. Altieri and Schmidt (1986a) conducted comparative studies on the ecology of arthropod communities in four ecologically different dry-farmed apple orchards: (1) an "abandoned" orchard not managed or disturbed for twenty-five years, (2) two "organic" (not sprayed with synthetic pesticides) orchards, one clean cultivated and the other with a mixed grass-legume cover crop, and (3) a "commercially" managed orchard (clean cultivated and subjected to chemical fertilizer and pesticide treatments). These ecologically different orchards constitute a "cultural evolution continuum." In the abandoned orchard, stable relationships between arthropods and the local vegetation have developed, probably because they are not disturbed. In the commercial orchard, high-energy inputs are substituted for some plant-insect interactions. The organic orchards combine characteristics of both systems. All orchards have at least one bordering edge with multilayered communities of wild vegetation.

Colonization of Orchards

The magnitude of the exchange of predators and parasitic Hymenoptera between the orchards and the wild vegetation edges are shown in Figure 7.7. Except for Syrphidae, considerably more individuals moved from edge to orchard in unsprayed (organic) than in sprayed orchards. Little exchange seemed to take place between abandoned orchards and woodlands. Figure 7.8 shows the temporal dynamics of colonization from the hedges to the sprayed and organic orchards by Coccinellidae. D-Vac samples taken from the shrub and herb layer of the edges revealed that edges of the organic orchards supported considerably more natural enemies than the edge of the sprayed orchard. In the early season, considerably more aphids invaded the sprayed orchards than the organic and abandoned orchards.

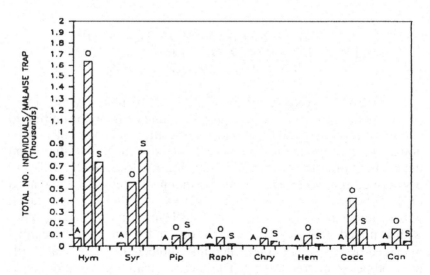

FIGURE 7.7. Seasonal mean number of natural enemies (Hym = parasitic Hymenoptera; Syr = Syrphidae adults; Pip = Pipinculidae; Raph = Raphididae; Chry = Chrysopidae; Hem = Hemiptera; Cocc = Coccinellidae; and Can = Cantharidae) caught in malaise traps in the interface between apple orchards (A = abandoned, O = organic, and S = sprayed) and wild vegetation edges in northern California (after Altieri and Schmidt, 1986a).

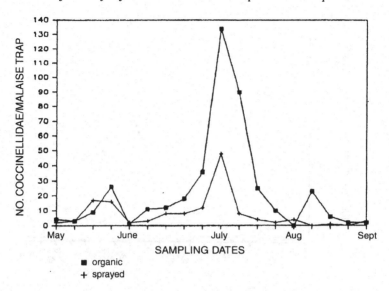

FIGURE 7.8. Mean number of adult Coccinellidae caught in malaise traps placed in the interface of woodlands and sprayed and organic apple orchards (after Altieri and Schmidt, 1986a).

Ground-Dwelling Predators

The ant species collected in the edges of the managed orchards also were found in the borders and centers, suggesting that part of the ant communities living in the wild margins colonized the planted orchards. Comparisons of mean ant abundance between edge, border, and center revealed major within-site differences. Throughout the season, more ants were caught in the borders or vegetation-free centers (Figure 7.9). Abundance gradually declined from the edge to the center of the orchards. In the abandoned orchard, however, ant species comparition and patterns of ant abundance were more or less uniform from edge to center from mid-May on. These trends can be explained by the structural similarities of the centers and edges. The vegetational complexity of the center of the abandoned system was similar to the edge, unlike the centers of the clean-cultivated systems that lacked the diversity of grasses and herbs characteristic of the edges.

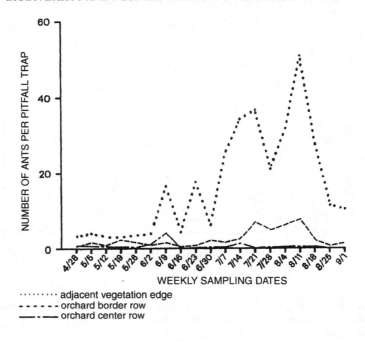

FIGURE 7.9. Mean number of ants caught in pitfall traps placed in the center and border tree rows of an organic apple orchard and in the adjacent vegetation edge (after Altieri and Schmidt, 1986a).

More ants were found in the wild vegetation edges of the managed orchards than in the edges of the abandoned orchards. Ants were most abundant in the abandoned edge early in the season, thereafter ant catches declined but then stabilized. Conversely, ant catches in the sprayed orchard's edge gradually increased and surpassed the abundance levels of ants in the organic edge from July on.

Spider catches were significantly higher in the edges of the abandoned orchards than in the organic or sprayed orchard edges. From May 15 through July 15, more spiders were caught in the organic edge than in the sprayed edge. Carabidae behaved differently than ants or spiders. Pitfall catches were higher in the center and border rows of apple trees than in the edges. This seems to be a normal pattern, as many Carabidae (i.e., *Agonum dorsale*) exhibit seasonal migration between field and edge.

Predation Pressure

Predation pressure (mainly by ants), as measured by the removal rates of potato tuberworm *(Phthorimaea operculella)* larvae from cardboard sheets placed on the orchard floor, was highest in the center of the abandoned orchard, followed by the organic orchards and the sprayed orchard (Table 7.2). Predation was greater in the center of the organic-cover orchard than in the center of the clean-cultivated one. On average, significantly more larvae were removed from the edges of the clean-cultivated organic and sprayed orchards than in the

TABLE 7.2. Removal of Potato Tuberworm Larvae, *Phthorimaea operculella* (Placed on the Ground), and of Mediterranean Flour Moth Eggs, *Anagasta kuehniella* (Placed on the Trees), by Predators in the Centers and Edges of Various Northern California Apple Orchards

Orchard System	% Eggs Removed[1]	% Larvae Removed[2]
Sprayed		
Center	21.0 ± 5.2	17.0 ± 6.2
Border	26.0 ± 8.0	
Edge	33.0 ± 10.2	32.7 ± 7.6
Organic		
Center	25.0 ± 8.0	61.5 ± 12.3
Border	34.1 ± 12.1	
Edge	43.1 ± 10.2	70.5 ± 13.9
Abandoned		
Center	38.0 ± 9.2	86.9 ± 14.5
Border	42.0 ± 10.7	
Edge	36.0 ± 4.2	86.4 ± 15.2

Source: after Altieri and Schmidt, 1986a.

[1]Predation pressure on the eggs was estimated on four occasions by hanging twenty-five 8.5 × 11.0 cm paper cards (with fifty moth eggs each) from the branches of each of five trees in the center, border, and edge of each orchard. [2]Means of three sampling dates. Larvae removal data were obtained by placing on the ground forty 22 × 22 cm cardboard sheets (twenty in the center and twenty on the edge) each containing twenty glued fourth instar larvae. Predation pressure was measured by determining removal of larvae in an eighteen-hour period.

centers of these same orchards. Removal rates, however, were similar in the wild vegetation edges and centers of the organic-cover and abandoned orchards. Cultivation and insecticide application probably disrupted ant communities in the centers of the clean-cultivated and sprayed orchards, confining ant foraging to the edges.

On the trees, predation of artificially placed *Anagasta kuehniella* eggs was consistently higher in the edges than in the border or center trees of the managed orchards. No differences in predation were observed between edge and center in the abandoned system. A clear gradient in predation pressure was observed in the centers of the orchards, declining from abandoned to sprayed.

MANIPULATING CROP-FIELD BORDER VEGETATION

It is clear then from these studies that the potential for pest regulation by certain natural enemies may be related to our ability to exercise some degree of control over the habitats surrounding crop fields. Could the herbivore-predator assemblages of an agroecosystem be manipulated by changing the vegetational composition and other features of surrounding edges and habitats? Some work conducted in England in recent years is providing some key information for management of field boundaries to increase natural enemy abundance and efficiency.

A strategy implemented by the UK Game Conservancy Trust and researchers of Southampton University consists of experimentally reducing field size by creating new, nonwoody, predator overwintering refuges. Raised banks (created by careful plowing) have been sown with grasses such as *Lolium, Dactylis, Agrostis,* and *Holcus.* Simply switching off the herbicide spray during normal cereal operations (plus the creation of adjacent sterile strips) will effectively provide new refuges hundreds of meters from existing orthodox field boundaries. Small (about 10 ha) and large (about 40 ha) fields have been used, and "peninsular" boundaries have been created which will reach the respective field center. This type of bank can also be sown with pollen-bearing and nectar-bearing plants in order to attract Hymenoptera and Syriphidae.

Workers at the Game Conservancy Trust have designed simple "minihedgerows," which can be accommodated at field margins within the width occupied by simple post-and-wire-strand fencing. These minihedgerows are nothing more complicated than a raised narrow strip, planted with suitable vegetation. The grasses *Dactylis glomerata* and *Holcus lanatus* seem particularly suitable plants for the beetles, but there is also scope for sowing flowering plants as adult food sources for other natural enemies. A further development is that such "predator conservation strips," running parallel with the crop rows (these banks are 0.4 m high, 1.5 m wide, and 300-400 m long and across the centers of the field), can be drilled at intervals across the crop to enhance natural enemy populations over the whole field area (Thomas and Wratten, 1990). These strips can be created afresh each year if the farmer wishes to change the direction of plowing. Research showed that high predator densities (nearly 1,500 predators/m^2) could be achieved in two years. An economic evaluation has shown that gains from the increased predator efficiency could more than repay the costs of labor and the expected 0.5 percent loss in crop yield, which together are less than any aphid-induced losses of about 5 percent or spraying costs equivalent to about 2.5 percent of crop yield. Research in Sweden by Chiverton (1989) showed that increased densities of cereal aphid predators (caradids such as *Bembidion lampros,* rove beetles of the genus *Tachyporus,* and several species of linyphiid spiders) were found overwintering on grassy banks, starting as soon as one year after bank establishment.

As suggested by studies in northern Florida, the species composition of weed communities of uncultivated land surrounding crop fields can be modified by plowing the land at different times of the year (Altieri and Whitcomb, 1979a). It was found that by increasing certain weeds experimentally, the diversity and numbers of herbivorous and predaceous insects associated with these weeds increased. Coccinellids were most abundant in plots plowed in December because these treatments enhanced the abundance of goldenrod (*Solidago* sp.) and Mexican tea *(Chenopodium ambrosioides),* which in turn provided suitable food (aphids and other herbivores) and habitat for the coccinellids and other predators (Figure 7.10). Hence, in this case, the manipulation of a particular predator colonizing cornfields depended upon the type and abundance of vegetation present around the fields as determined by the time of plowing.

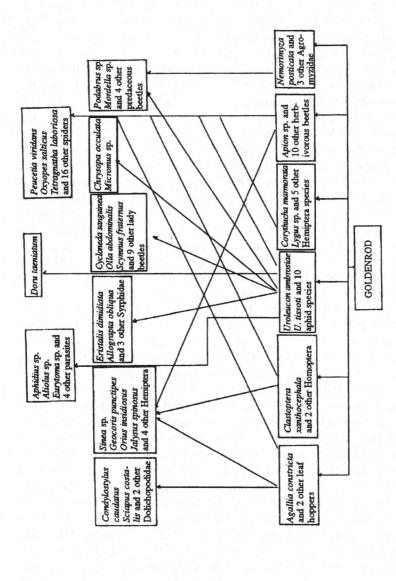

FIGURE 7.10. Food web formed by the major arthropod species associated with goldenrod (*Solidago altissima*) in north Florida (after Altieri and Whitcomb, 1979).

CASE STUDY 2: BIOLOGICAL CORRIDORS
IN VINEYARDS

As mentioned, research by Kido et al. (1981) established that French prunes *(Prunus domestica)* adjacent to vineyards could also serve as overwintering sites for the parasitic wasp *A. epos,* and Murphy, Rosenheim, and Granett (1996) detected higher grape leafhopper parasitism in grape vineyards with adjacent prune tree refuges than in vineyards lacking refuges. Corbett and Rosenheim (1996), however, determined that the effect of prune refuges was limited to few vine rows downwind, and *A. epos* exhibited a gradual decline in vineyards with increasing distance from the refuge. This finding poses an important limitation to the use of prune trees, as the colonization of grapes by *A. epos* is limited to field borders leaving the central rows of the vineyard void of biological-control protection.

To overcome this limitation, Nicholls, Parrella, and Altieri (2000) tested whether an established vegetational corridor enhanced movement of beneficial insects beyond the "normal area of influence" of adjacent habitats or refuges. The study was conducted in northern California in 1996 and 1997 and involved two adjacent vineyards surrounded on the north side by riparian forest vegetation; the main difference between the two vineyards is that vineyard A was penetrated and dissected by a five-meter-wide and 300-meter-long vegetational corridor composed of sixty-five different species of flowering plants. Vineyard B had no corridor.

In both years in vineyard A, adult leafhoppers exhibited a clear density gradient reaching lowest numbers in vine rows near the corridor and forest and increasing in numbers toward the center of the field, away from the adjacent vegetation. The highest concentration of leafhoppers occurred after the first twenty to twenty-five rows (30 to 40 m) downwind from the corridor. Such a gradient was not apparent in vineyard B, where the lack of the corridor resulted in a uniform dispersal pattern of leafhoppers (Figures 7.11 and 7.12). Nymphal populations behaved similarly, reaching the highest numbers in the center rows of block A in both years.

Generalist predators in the families Coccinellidae, Chrysopidae, Nabidae, and Syrphidae exhibited a density gradient in vineyard A, indicating that the abundance and spatial distribution of these insects was influenced by the presence of the corridor which channeled dis-

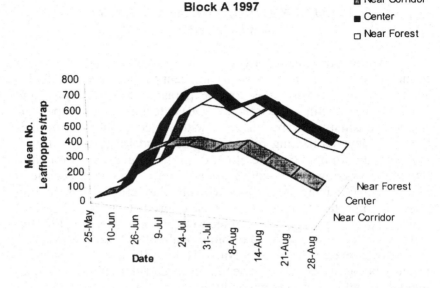

FIGURE 7.11. Seasonal patterns (numbers per yellow sticky trap) of adult leafhopper *E. elegantula* in block A, as influenced by the presence of the corridor (*P* < 0.05; Mann-Whitney U-test) (Hopland, California, 1997) (after Nicholls, Parrella, and Altieri, 2001).

persal of the insects into adjacent vines (Figures 7.13 and 7.14). Predators were more homogeneously (but reaching lower overall abundance) distributed in vineyard B, as no differences in spatial pattern in predator catches was observed between bare edge and central rows.

Anagrus epos colonized vineyards from the corridor and forest throughout the sampling area, exhibiting higher densities in late July and throughout August of both years in the central vineyard rows where leafhoppers were most abundant. By following the abundance patterns of leafhoppers, the *Anagrus* wasp did not display the distributional response exhibited by predators. For this reason, these researchers concluded that predator enhancement near the vegetational corridor explained the lower populations of leafhoppers and thrips in the first twenty-five rows. Such successful impact of predators can be assumed because fewer adults and nymphs of leafhoppers and thrips

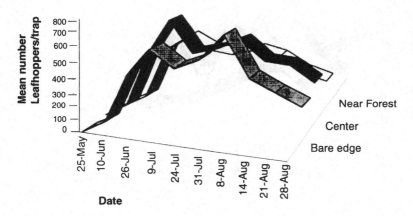

FIGURE 7.12. Seasonal patterns (numbers per yellow sticky trap) of adult leafhopper *E. elegantula* in block B without the presence of the corridor but with an adjacent forest ($P < 0.05$; Mann-Whitney U-test) (Hopland, California, 1997) (after Nicholls, Parrella, and Altieri, 2001).

were caught near the corridor than in the middle of the vineyards. Overall abundance of predators was higher in vineyard A than B throughout the season (Figure 7.15). The corridor provided a constant supply of alternative food for predators, effectively decoupling predators from a strict dependence on grape herbivores and avoiding a delayed colonization of the vineyard. This complex of predators continuously circulated into the vineyard interstices, establishing a set of trophic interactions leading to lower numbers of leafhoppers and thrips in the border rows of the vineyard.

Findings from this study also suggest that the creation of corridors across vineyards can serve as a key strategy to allow natural enemies emerging from riparian forests to disperse over large areas of otherwise monoculture systems. Such corridors should be composed of locally adapted plant species exhibiting sequential flowering periods, which attract and harbor an abundant diversity of predators and

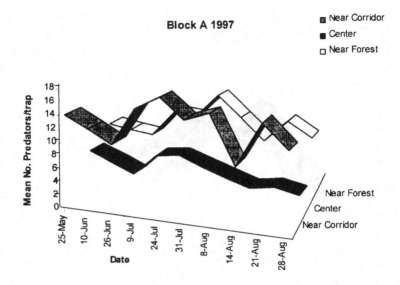

FIGURE 7.13. Seasonal patterns of predator catches (numbers per yellow sticky trap) in block A, as influenced by the presence of forest edge and the corridor ($P < 0.05$; Mann-Whitney U-test) (Hopland, California, 1997) (after Nicholls, Parrella, and Altieri, 2001).

parasitoids and increase biodiversity. Thus, these corridors or strips, which may link various crop fields and riparian forest remnants, can create a network that would allow many species of beneficial insects to disperse throughout whole agricultural regions transcending farm boundaries (Baudry, 1984).

CASE STUDY 3: STRIP MANAGEMENT TO AUGMENT PREDATORS

As a way to enhance predator abundance in cereal fields, researchers in Switzerland introduced vegetation edges as successional strips into the field. One 8 ha winter cereal field was subdivided by five wide-strips leaving cereal spaces of 12, 24, and 36 m between the strips (Lys and Nentwig, 1992). Significantly higher recapture rates, indicating higher predator activity, were found in the strip-managed

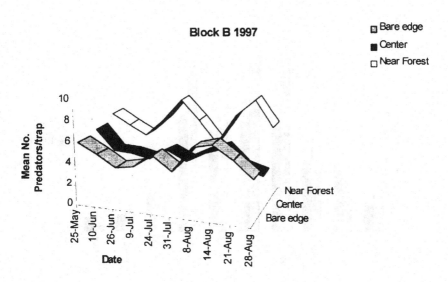

FIGURE 7.14. Seasonal patterns of predator catches (numbers per yellow sticky trap) in block B without the corridor but with an adjacent forest ($P < 0.05$; Mann-Whitney U-test) (Hopland, California, 1997) (after Nicholls, Parrella, and Altieri, 2001).

area than in the control area, especially for carabid beetles such as *Poecilus cupreus, Carabus granulatus,* and *Pterostichus melanarius.* Several observations led to the conclusion that this higher activity was generally due to a prolongation of the reproductive period in the strip-managed area.

Besides the marked increase in activity and density, a large increase in the diversity of ground beetle species was observed. The most marked increase in number of species was found in the first year. The vegetation structure of the cereal field was enriched following use of weed strips. After three years of research the authors concluded that weed strips offer not only higher food availability but also more suitable overwintering sites. In addition, these weed strips offer refuges during field disturbance or during unfavorable climatic conditions, such as droughts. Weed strips increase the chance of survival of many carabid species in arable ecosystems, thus counteracting the faunal impoverishment trends promoted by monocultures. Nentwig

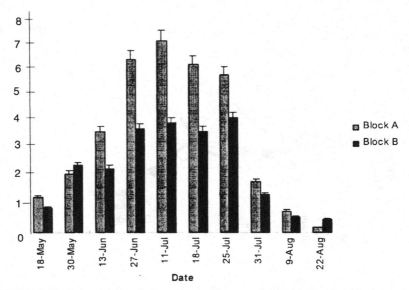

FIGURE 7.15. Comparison of abundance of generalist predators (numbers per yellow sticky trap) between block A (with a corridor) and block B (without a corridor) ($P < 0.05$, Wilcoxon's signed rank test) (Hopland, California, 1996) (after Nicholls, Parrella, and Altieri, 2001).

(1998) found similar effects with 3- to 9-m-wide sown weed strips dividing large fields into small parts so that the distance between strips does not exceed 50 to 100 m. A favorite plant to be used as strips within or around fields is *Phacelia tanacetifolia* (Holland and Thomas, 1996).

In reviewing these studies Corbett and Plant (1993) argued for the need to develop a mechanistic framework to evaluate and predict the response of natural enemies to vegetational arrangements in agroecosystems. Using a hypothetical field with 10 m wide strips interplanted at 100 m intervals (Figure 7.16), they assumed that these strips are used solely as an overwintering refuge by three natural enemy species: (1) a predatory mite having very low mobility (a diffusion coefficient of 1 m^2/day); (2) a predatory coccinellid beetle having moderate mobility (10 m^2/day); and (3) a highly mobile parasitoid (100 m^2/day). Once the crop has germinated, the strips do not provide resources in any greater abundance than the crop, nor do they provide a

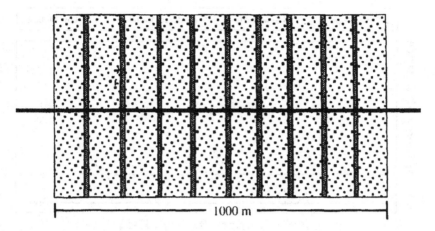

FIGURE 7.16. Diagram of hypothetical diversified agroecosystem. Interplanted strips are placed 100 m apart within a crop. The model predicts natural enemy abundance along a transect through the field (after Corbett, 1998).

more favorable physical habitat. The mobility (i.e., the probability of making a move in a given time period) is therefore the same in the strips as it is in the crop. Natural enemies overwinter in the strips at a density of ten individuals per square meter.

The spatial patterns in abundance predicted by the model for these three hypothetical natural enemies are illustrated in Figure 7.17. The model predicts that the natural enemies will spread from the strips, resulting in higher abundance in the crop than would have occurred in a crop monoculture. The distance to which they are enhanced varies substantially, however. For the predatory mite, enhancement is confined to the region immediately adjacent to interplanted strips, producing a steep gradient in density with increasing distance. The highly mobile parasitoid, on the other hand, is enhanced throughout the crop—there is no spatial pattern to suggest that strips influenced abundance. As a result, natural enemies with low mobility exhibit no enhancement beyond 20 m from strips, while more mobile natural enemies are enhanced fourfold.

Corbett and Plant (1993) proposed a second scenario using the same field and natural enemies. In this scenario, however, the interplanted vegetational zones are not overwintering refuges: natural enemies must colonize the agroecosystem from external sources. The

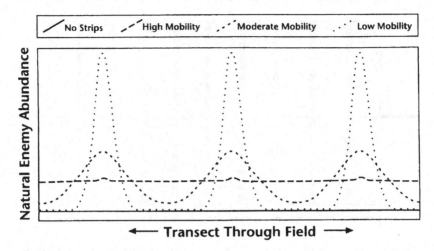

FIGURE 7.17. Spatial patterns predicted by model for field with interplanted strips that serve solely as an overwintering refuge. Peak abundance occurs at interplanted strips. Patterns are shown for hypothetical natural enemies of three different mobilities: low mobility (1 m²/day), moderate mobility (10 m²/day), high mobility (100 m²/day) (after Corbett, 1998).

strips do, however, provide more resources than the crop. Therefore, the probability of making a move in a given time period is lower in the strips than in the crop. The resources in interplantings are assumed to be "substitutable," to some degree, for resources that occur in the crop. They could be either (1) alternate prey or hosts ("supplementary" resources) or (2) floral resources that are an imperfect substitute for the preferred host but that benefit the natural enemy when available ("complementary" resources).

The model predicts that natural enemy mobility would dramatically affect the observed enhancement due to increased diversity (Figure 7.18). Natural enemies that are highly mobile would show little enhancement when plots are 50 m wide because predators are dispersing among all plots in the experimental field. The observed enhancement increases with plot size since as strips are farther apart their effect becomes more detectable. However, even plots 200 m in size do not detect the enhancement that would occur in a diversified, commercial-scale field. For less mobile natural enemies, enhancement is observed when plots are small, but the observed enhancement

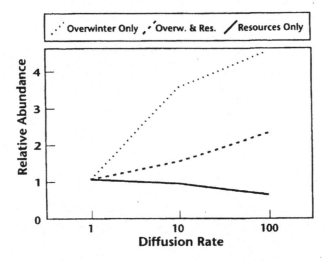

FIGURE 7.18. Effect of mobility on the abundance of natural enemies on crop vegetation in a diversified agroecosystem. "Relative Abundance" is the ratio of natural enemy abundance predicted for the diversified system to that predicted for a crop monoculture. Relative abundance is calculated only for crop vegetation more than 20 m from interplantings. Effect of diversification is shown for three different situations: where the interplantings act solely as an overwintering refuge; where they provide additional food resources but no overwintering sites; and where interplantings provide both (after Corbett and Plant, 1993).

decreases with increasing plot size. This is because such predators are enhanced only in the area adjoining strips.

The model also predicts that the abundance of the three natural enemies is higher in the interplanted strips than in the crop vegetation. This accumulation of natural enemies in the strips is due to the lower tendency for movement there and results in the strips acting as a sink for the natural enemies. The parasitoid exhibits a spatially uniform density in the crop vegetation and the greatest accumulation in the strips. This sink effect results in abundance on the crop that is 60 percent of what it would be in an undiversified field. The other natural enemies exhibit some spatial patterning within the crop and a milder sink effect.

Chapter 8

The Dynamics of Insect Pests
in Agroforestry Systems

Agroforestry is an intensive land-management system that combines trees and/or shrubs with crops and/or livestock (Nair, 1993). Many of the benefits of agroforestry are derived from the increased diversity of these systems compared to corresponding monocultures of crops or trees. Despite the fact that little research has been conducted on pest interactions within agroforestry systems, agroforestry has been recommended to reduce pest outbreaks usually associated with monocultures. Although the effects of various agroforestry designs on pest populations can be of a varied nature (microclimatic, nutritional, natural enemies, etc.), regulating factors do not act in isolation from one another.

The few reviews on pest management in agroforestry (Schroth et al., 2000; Rao, Singh, and Day, 2000) expect that high plant diversity protects agroforestry systems to some extent from pest and disease outbreaks. These authors use the same theories advanced by agroecologists to explain lower pest levels in polycultural agroecosystems as discussed in Chapter 3. These authors also caution that the use of high plant diversity as a strategy to reduce pest and disease risks in agroforestry systems involves considerable technical and economic difficulties. Whereas a farmer is free to cultivate crops either on separate fields or in association, the choice of the crops themselves (and thus the overall crop diversity of the farm) is strongly influenced by the availability of markets for the respective products and the needs of the household. The selection of timber and fruit trees also has to respect local market conditions, although more freedom of choice may exist for "service" trees, for example, trees grown for biomass, shade, or wind protection.

THE EFFECTS OF TREES IN AGROFORESTRY
ON INSECT PESTS

The deliberate association of trees with agronomic crops can result in insect-management benefits due to the structural complexity and permanence of trees and to their modification of microclimates and plant apparency within the production area.

Individual plants in annual cropping systems are usually highly synchronized in their phenology and are short lived. The lack of temporal continuity is a problem for natural enemies because prey availability is limited to short periods of time and refugia, and other resources are not available consistently. The addition of trees of variable phenologies or diverse age structure through staggered planting can provide refuge and a more constant nutritional supply to natural enemies because resource availability through time is increased. Trees can also provide alternate hosts to natural enemies, as in the case of the planting of prune trees adjacent to grape vineyards to support overwintering populations of the parasitoid *A. epos* (Murphy et al., 1998).

Shade from trees may markedly reduce pest density in understory intercrops. Hedgerows or windbreaks of trees have a dramatic influence on microclimate; almost all microclimate variables (heat input, wind speed, soil desiccation, and temperature) are modified downwind of a hedgerow. Tall intercrops or thick ground covers can also alter the reflectivity, temperature, and evapotranspiration of shaded plants or at the soil surface, which in turn could affect insects that colonize according to "background" color or that are adapted to specific microclimatological ranges (Cromartie, 1981). Both immature and adult insect growth rates, feeding rates, and survival can be dramatically affected by changes in moisture and temperature (Perrin, 1977).

The effect of shade on pests and diseases in agroforestry has been studied quite intensively in cocoa and coffee systems undergoing transformation from traditionally shaded crop species to management in unshaded conditions. In cocoa plantations, insufficient overhead shade favors the development of numerous herbivorous insect species, including thrips *(Selenothrips rubrocinctus)* and mirids *(Sahlbergella* sp., *Distantiella* sp., etc.). Even in shaded plantations, these insects concentrate in spots where the shade trees have been destroyed, such as by wind (Beer et al., 1997). Bigger (1981) found an increase in the numbers of Lepidoptera, Homoptera, Orthoptera, and the mirid *Sahlbergella singularis* and a decrease in the number of Diptera and parasitic Hymenoptera from the shaded toward the unshaded part of a cocoa plantation in Ghana.

In coffee, the effect of shade on insect pests is less clear than in cocoa, as the leaf miner *(Leucoptera meyricki)* is reduced by shade, whereas the coffee berry borer *(Hypothenemus hampei)* may increase under shade. Similarly, unshaded tea suffers more from attack by thrips and mites, such as the red spider mite *(Oligonychus coffeae)* and the pink mite *(Acaphylla theae)*, whereas heavily shaded and moist plantations are more damaged by mirids (*Helopeltis* spp.) (Guharay et al., 2000).

Although in Central America coffee berry borer appears to perform equally well in open sun and managed shade, naturally occurring *Beauveria bassiana* (an entomopathogenic fungus) multiplies and spreads more quickly with greater humidity, and fungus applications should coincide with peaks in rainfall (Guharay et al., 2000). After a study of how the microclimate created by multistrata shade management affected herbivores, diseases, weeds, and yields in Central America coffee plantations, Staver and colleagues (2001) defined the conditions for minimum expression of the pest complex. For a low-elevation dry coffee zone, shade should be managed between 35 to 65 percent, as shade promotes leaf retention in the dry season and reduces *Cercospora coffeicola,* weeds, and *Planococcus citri.* Obviously, the optimum shade conditions for pest suppression differs with climate, altitude, and soils. The selection of tree species and associations, density, and spatial arrangements, as well as shade-management regimes, are critical considerations for shade-strata design (Figure 8.1).

The complete elimination of shade trees may have an enormous impact on the diversity and density of ants. Studying the ant community in a gradient of coffee plantations going from plantations with high density of shade to shadeless plantations, Perfecto and Vandermeer (1996) reported a significant decrease in ant diversity. Although the relationship between ant diversity and pest control is not well understood, we can speculate that a diverse ant community can offer more safeguards against pest outbreaks than a community dominated by just a few species. In Colombia, preliminary reports point to lower levels of the coffee borer, the main coffee pest in the region, in shaded coffee plantations. There are more indications that a non-

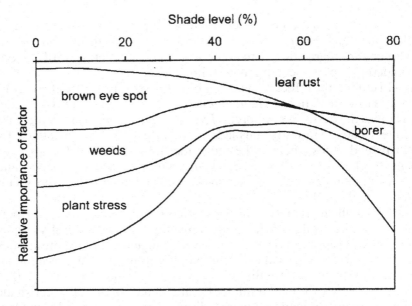

FIGURE 8.1. Conceptual graph depicting the relative importance of yield-reducing factors in a low, dry coffee zone in Nicaragua. Effects are shown to be additive with the effect of each successive pest represented by the area between the lines. The lowest line indicates the accumulated potential for yield reduction at different shade levels. Since the y-axis is negative, the range of least yield reduction is 35 to 65 percent (after Guharay et al., 2000).

dominant small ant species is responsible for the control. Apparently, this species does not live in unshaded plantations. Cocoa is another crop that is traditionally cultivated under shade trees. The ant species that have been so successful in controlling pests in cocoa are all species that flourish under shaded conditions.

One of the most obvious consequences of pruning or shade elimination, with regard to the ant community, is the change in microclimatic conditions. In particular, microclimate becomes more variable with more extreme levels of humidity and temperature. A recent study documented changes in the composition of the ant community with shade and leaf-litter manipulation similar to those that occur after plowing (Perfecto and Vandermeer, 1996).

Chemical cues used by herbivores may be altered in an agroforestry system. Trees may exhibit a dramatically different chemical profile than annual herbaceous plants intercropped in the system, masking or lessening the impact of the chemical profile produced by the annual crop. Several studies have demonstrated olfactory deterrence as a factor in decreasing arthropod abundance (Risch, 1981).

The attractiveness of a plant species for the pests of another species can be usefully employed in agroforestry associations in the form of trap crops which concentrate the pests or disease vectors, a place where they cause less damage or can be more easily neutralized (e.g., by spraying or collecting). Such trap crops are an interesting option when they attract pests from the primary crop within the field (local attraction) but not when they attract pests from areas outside the field (regional attraction). Nascimento, Mesquita, and Caldas (1986) demonstrated the strong attraction of the *Citrus* pest *Cratosomus flavofasciatus* by the small tree *Cordia verbenacea* in Bahia, Brazil, and recommended the inclusion of this tree at distances of 100 to 150 m in *Citrus* orchards. They speculated that pests of several other fruit crops could similarly be trapped by this tree species.

In certain agroforestry systems, such as alley cropping or systems with perennial crops and leguminous shade trees, relatively large quantities of N-rich biomass may be applied to crops. In cases of luxury consumption of N, this may result in reduced pest resistance of the crops. The reproduction and abundance of several insect pests, especially Homoptera, are stimulated by the high concentration of free nitrogen in the crop's foliage.

In studies assessing the effects of nine hedgerow species on the abundance of major insect pests of beans and maize, and associated predatory/parasitic anthropods, Girma, Rao and Sithanantham (2000) found that beanfly (*Ophiomyia* spp.) infestation was significantly higher in the presence of hedgerows (35 percent) than in their absence (25 percent). Hedgerows did not influence aphid *(Aphis fabae)* infestation of beans. In contrast, maize associated with hedgerows experienced significantly lower stalk borer (*Busseola fusca* and *Chilo* spp.) and aphid *(Rhopalosiphum maidis)* infestations than pure maize, the margin of difference being 13 percent and 11 percent respectively for the two pests. Ladybird beetles closely followed their prey, aphids, with significantly higher catches in sole-cropped plants than in hedgerow plots and away from hedgerows. Activity of wasps was significantly greater close to hedgerows than away from them. Spider catches during maize season were 77 percent greater in the presence of hedgerows than in their absence, but catches during other seasons were similar between the two cropping systems.

In one of the few studies of the influence of temperate agroforestry practices on beneficial arthropods, Peng and colleagues (1993) confirmed the increase in insect diversity and improved natural enemy abundance in an alley cropping system over that of a monoculture crop system. Their study examined arthropod diversity in control plots sown to peas (*Pisum sativum* var. Sotara) versus peas intercropped with four tree species (walnut, ash, sycamore, and cherry) and hazel bushes. They found greater arthropod abundance in the alley-cropped plots compared to the control plots, and natural enemies were more abundant in the tree lines and alleys than in the controls. The authors attributed the increase in natural enemies to the greater availability of overwintering sites and shelter in the agroforestry system. In fact, Stamps and Linit (1997) argue that agroforestry holds promise for increasing insect diversity and reducing pest problems because the combination of trees and crops provides greater niche diversity and complexity in both time and space than does polyculture of annual crops.

DESIGNING NATURAL SUCCESSIONAL ANALOG AGROFORESTRY SYSTEMS

At the heart of the agroecology strategy is the idea that an agroecosystem should mimic the diversity and functioning of local ecosystems, thus exhibiting tight nutrient cycling, complex structure, and enhanced biodiversity. The expectation is that such agricultural mimics, like their natural models, can be productive, pest resistant, and conservative of nutrients and biodiversity. Ewel (1986) argues that natural plant communities have several traits (pest suppression among them) that would be desirable to incorporate into agroecosystems. In order to test this idea, he and others (Ewel et al., 1982) studied productivity, growth, resilience, and resource-utilization characteristics of tropical successional plant communities. These researchers contend that desirable ecological patterns of natural communities should be incorporated into agriculture by designing cropping systems that mimic the structural and functional aspects of secondary succession. Thus, the prevalent coevolved natural secondary plant associations of an area should provide the model for the design of multispecies crop mixtures (Soule and Piper, 1992).

This succession analog method requires a detailed description of a natural ecosystem in a specific environment and the botanical characterization of all potential crop components. When this information is available, the first step is to find crop plants that are structurally and functionally similar to the plants of the natural ecosystem. The spatial and chronological arrangement of the plants in the natural ecosystem are then used to design an analogous crop system (Hart, 1980). In Costa Rica, researchers conducted spatial and temporal replacements of wild species with botanically, structurally, and ecologically similar cultivars. Thus, successional members of the natural system such as *Heliconia* spp., cucurbitaceous vines, *Ipomoea* spp., legume vines, shrubs, grasses, and small trees were replaced by plantain, squash varieties, and yams. By years two and three, fast-growing tree crops (Brazil nuts, peach, palm, rosewood) may form an additional stratum, thus maintaining continuous crop cover, avoiding site degradation and nutrient leaching, and providing crop yields throughout the year (Ewel, 1986).

Under a scheme of managed succession, natural successional stages are mimicked by intentionally introducing plants, animals, practices, and inputs that promote the development of interactions and connections between component parts of the agroecoystem. Plant species (both crop and noncrop) are planted that capture and retain nutrients in the system and promote good soil development. These plants include legumes, with their nitrogen-fixing bacteria, and plants with phosphorus-trapping mycorrhizae. As the system develops, increasing diversity, food-web complexity, and level of mutualistic interactions all lead to more effective feedback mechanisms for pest and disease management. The emphasis during the development process is on building a complex and integrated agroecosystem with less dependence on external inputs.

There are many ways that a farmer, beginning with a recently cultivated field of bare soil, can allow successional development to proceed beyond the early stages. One general model is to begin with an annual monoculture and progress to a perennial tree crop system.

1. The farmer begins by planting a single annual crop that grows rapidly, captures soil nutrients, gives an early yield, and acts as a pioneer species in the developmental process.
2. As a next step (or instead of the previous one), the farmer can plant a polyculture of annuals that represent different components of the pioneer stage. The species would differ in their nutrient needs, attract different insects, have different rooting depths, and return a different proportion of their biomass to the soil. One might be a nitrogen-fixing legume. All of these early species would contribute to the initiation of the recovery process, and they would modify the environment so that noncrop plants and animals—especially the macroorganisms and microorganisms necessary for developing the soil ecosystem—can also begin to colonize.
3. Following the initial stage of development, short-lived perennial crops can be introduced. Taking advantage of the soil cover created by the pioneer crops, these species can diversify the agroecosystem in important ecological aspects. Deeper root systems, more organic matter stored in standing biomass, and greater habitat and microclimate diversity all combine to advance the successional development of the agroecosystem.
4. Once soil conditions improve sufficiently, the ground is prepared for planting longer-lived perennials, especially orchard or tree crops, with annual and short-lived perennial crops maintained in the areas between them. While the trees are in their early growth, they have limited impact on the environment around them. At the same time, they benefit from having annual crops around them, because in the early stages of growth they are often more susceptible to interference from aggressive weedy species that would otherwise occupy the area.
5. As the tree crops develop, the space between them can continue to be managed with annuals and short-lived perennials.
6. Eventually, once the trees reach full development, the end point in the developmental process is achieved. This last stage is dominated by woody plants, which are key to the site-restoring growers of fallow vegetation because of their deep, permanent root systems.

Once a successionally developed agroecosystem has been created, the problem becomes how to manage it. The farmer has three basic options:

- Return the entire system to the initial stages of succession by introducing a major disturbance, such as clear-cutting the trees in the perennial system. Many of the ecological advantages that have been achieved will be lost and the process must begin anew.
- Maintain the system as a tree-crop-based agroecosystem.
- Reintroduce disturbance into the agroecosystem in a controlled and localized manner, taking advantage of the dynamics that such patchiness introduces into an ecosystem. Small areas in the system can be cleared, returning those areas to earlier stages in succession, and allowing a return to the planting of annual or short-lived crops. If care is taken in the disturbance process, the belowground ecosystem can be kept at a later stage of development, whereas the aboveground system can be made up of highly productive species that are available for harvest removal.

According to Ewel (1999), the only region where it pays to imitate natural ecosystems, rather than struggle to impose simplicity through high inputs in ecosystems that are inherently complex, is the humid tropical lowlands. This area epitomizes environments of low abiotic stress but overwhelming biotic intricacy. The keys to agricultural success in this region are to (1) channel productivity into outputs of nutritional and economic importance, (2) maintain adequate vegetational diversity to compensate for losses in a system simple enough to be horticulturally manageable, (3) manage plants and herbivores to facilitate associational resistance, and (4) use perennial plants to maintain soil fertility, guard against erosion, and make full use of resources.

To many, the ecosystem-analog approach is the basis for the promotion of agroforestry systems, especially the construction of forest-like agroecosystems that imitate successional vegetation and exhibit low requirements for fertilizer, high use of available nutrients, and high protection from pests (Sanchez, 1995).

THE NEED FOR FURTHER RESEARCH

The effects of agroforestry plants (or techniques) on pests and diseases can be divided into biological (species-related) and physical effects of components (e.g., microclimate). The former is highly specific for certain plant-pest or plant-disease combinations and must be studied on a case-by-case basis. The latter are easier to generalize, but even they depend on the regional climatic conditions. Based on results from intercropping studies, agroforesters expect that agroforestry systems may provide opportunities to noticeably increase arthropod diversity and lower pest populations compared to the polyculture of annual crops or trees by themselves. However, more work is needed in specific areas of research such as studies of the differences in arthropod populations between agroforestry and traditional agronomic systems, research into the specific mechanisms behind enhancement of pest management with agroforestry practices, and basic research into the life histories of target pests and potential natural enemies. An understanding of what aspects of trees modify pest populations—shelter, food, or host resources for natural enemies, temporal continuity, microclimate alteration or apparency—should help in determining future agroforestry design practices (Rao, Singh, and Day, 2000).

Well-designed agroforestry techniques can reduce crop stress by providing the right amount of shade, reducing temperature extremes, sheltering off strong winds, and improving soil fertility, thereby improving the tolerance of crops against pest and disease damage, while at the same time influencing the development conditions for pest and disease organisms and their natural enemies. Poorly designed systems, on the other hand, may increase the susceptibility of crops to pests.

Chapter 9

Designing Pest-Stable Vegetationally Diverse Agroecosystems

MONOCULTURES AND THE FAILURE OF CONVENTIONAL PEST-CONTROL APPROACHES

Until about four decades ago, crop yields in U.S. agricultural systems depended on internal resources, recycling of organic matter, built-in biological control mechanisms, and rainfall patterns. Agricultural yields were modest but stable. Production was safeguarded by growing more than one crop or variety in space and time in a field as insurance against pest outbreaks or severe weather. Inputs of nitrogen were gained by rotating major field crops with legumes. In turn, rotations suppressed insects, weeds, and diseases by effectively breaking the life cycles of these pests. A typical corn-belt farmer grew corn rotated with several crops, including soybeans, and small-grain production was intrinsic to maintain livestock (USDA, 1973). Most of the labor was done by the family with occasional hired help, and no specialized equipment or services were purchased from off-farm sources (Altieri, 1995; Gliessman, 1999). In the developing world, small farmers developed even more complex and biodiverse farming systems guided by indigenous knowledge that has stood the test of time (Thrupp, 1997). In these types of farming systems, the link between agriculture and ecology was quite strong, and signs of environmental degradation were seldom evident.

As agricultural modernization progressed, however, the ecology-farming linkage was often broken as ecological principles were ignored and/or overridden. As profit rather than people's needs or environmental concerns shaped the modes of agricultural production, agribusiness interests and prevailing policies favored large farm size, specialized production, crop monocultures, and mechanization.

Today, monocultures have increased dramatically worldwide, mainly through the geographical expansion of land annually devoted to single crops. Thus, monoculture has implied the simplification of biodiversity, the end result being an artificial ecosystem requiring constant human intervention in the form of agrochemical inputs which, in addition to temporarily boosting yields, result in a number of undesirable environmental and social costs. As long as the structure of monocultures is maintained as the structural base of modern agricultural systems, pest problems will continue to be the result of a negative treadmill that reinforces itself (Figure 9.1). The exacerbation of

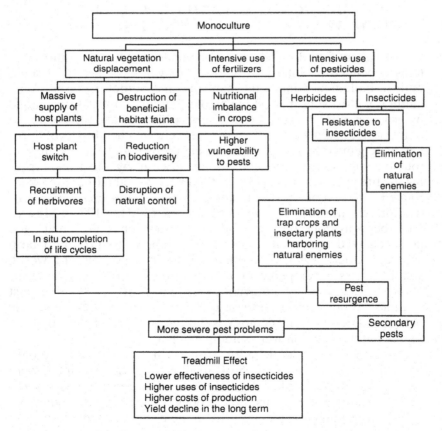

FIGURE 9.1. The ecological consequences of monoculture with special reference to pest problems and the agrochemical treadmill.

pest problems due to vegetational simplification has been the subject of this book, but pest problems can also result from excessive use of chemical fertilizers and pesticides (Phelan, Mason, and Stinner, 1995).

The substantial yield losses due to pests, about 20 to 30 percent for most crops (similar pest-loss levels were observed forty years ago) despite the increase in the use of pesticides (about 4.7 billion pounds of pesticides were used worldwide in 1995, 1.2 billion pounds in the United States alone) is a symptom of the environmental crisis affecting agriculture. Cultivated plants grown in genetically homogeneous monocultures do not possess the necessary ecological defense mechanisms to tolerate the impact of pest outbreaks. Modern agriculturists have selected crops mainly for high yields and high palatability, making them more susceptible to pests by sacrificing natural resistance for productivity. Consequently, as modern agricultural practices reduce or eliminate the resources and opportunities for natural enemies of pests, their numbers decline, decreasing the biological suppression of pests. Due to this lack of natural controls, an investment of about $40 billion in pesticide control is incurred yearly by U.S. farmers, which is estimated to save approximately $16 billion in U.S. crops. However, the indirect costs of pesticide use to the environment and public health have to be balanced against these benefits. Based on the available data, the environmental cost (impacts on wildlife, pollinators, natural enemies, fisheries, water, and development of resistance) and social costs (human poisonings and illnesses) of pesticide use reach about $8 billion each year (Conway and Pretty, 1991). What is worrisome is that pesticide use is still high and still rising in some cropping systems. Data from California show that from 1991 to 1995 pesticide use increased from 161 to 212 million pounds of active ingredient. This increase was not due to increases in planted acreage, as statewide crop acreage remained constant during this period. Much of the increase was due to intensification of some crops (grapes, strawberries) and higher uses of particularly toxic pesticides, many of which are linked to cancer (Liebman, 1997).

On the other hand, increasing evidence suggests that crops grown in chemically fertilized systems are more susceptible to pest attacks than crops grown in organic-rich and biologically active soils. Many studies suggest that the physiological susceptibility of crops and pathogens may be affected by the form of fertilizer used (organic versus chemical fertilizer). Studies documenting lower densities of several insect herbivores in low-input systems have partly attributed such reduction to a low nitrogen content in organically farmed crops (Altieri, 1995).

Given these findings, a major challenge for scientists and farmers advocating for more ecologically based pest management (EBPM) is to find strategies to overcome the ecological limits imposed by monocultures (lack of diversity and foliage nutritional imbalances), by converting such systems into diversified agroecosystems dependent on internal resources and aboveground and belowground synergism rather than on high external inputs. This chapter provides some ideas and principles of agroecosystem design that may lead to a more optimal biological pest regulation.

TOWARD SUSTAINABLE AGRICULTURE

The search for self-sustaining, low-input, diversified, and energy-efficient agricultural systems is now a major concern of many researchers, farmers, and policymakers worldwide. More sustainable food-production systems seek to make the best use of nature's goods and services while not damaging the environment (Altieri, 1995, 1999; Thrupp, 1997; Conway, 1994; Pretty, 1995, 1997; Pretty and Hine, 2000). This can be done by integrating natural processes such as nutrient cycling, nitrogen fixation, soil regeneration, and natural enemies of pests into food-production processes. A sustainable agriculture also minimizes the use of nonrenewable inputs (pesticides and fertilizers) that damage the environment or harm the health of farmers and consumers by encouraging regenerative and resource-conserving technologies. It makes better use of the knowledge and skills of farmers, improving their self-reliance. It also seeks to make productive use of people's capacities to work together to solve common management problems, such as pest problems at a regional level. Most people involved in the promotion of sustainable agriculture aim at creating a form of agriculture that maintains productivity in the long term by doing the following (Pretty, 1995; Altieri, 1995):

- Optimizing the use of locally available resources by combining the different components of the farm system, i.e., plants, ani-

mals, soil, water, climate, and people, so that they complement one another and have the greatest possible synergetic effects
- Reducing the use of off-farm, external, and nonrenewable inputs with the greatest potential to damage the environment or harm the health of farmers and consumers, and a more targeted use of the remaining inputs employed with a view to minimizing variable costs
- Relying mainly on resources within the agroecosystem by replacing external inputs with nutrient cycling, better conservation, and an expanded use of local resources
- Improving the match between cropping patterns and their productive potential and environmental constraints of climate and landscape to ensure long-term sustainability of current production levels
- Working to value and conserve biological diversity, both in the wild and in domesticated landscapes, and making optimal use of the biological and genetic potential of plant and animal species
- Taking full advantage of local knowledge and practices, including innovative approaches not yet fully understood by scientists although widely adopted by farmers

As it has been emphasized in this book, a key strategy in sustainable agriculture is to restore agricultural diversity of the agricultural landscape (Altieri, 1987). Diversity can be enhanced in time through crop rotations and sequences and in space in the form of cover crops, intercropping, agroforestry, crop-livestock mixtures, and manipulation of vegetation outside the crop area. Vegetation diversification not only results in pest regulation through restoration of natural control but also produces optimal nutrient recycling, soil conservation, energy conservation, and less dependence on external inputs.

REQUIREMENTS OF SUSTAINABLE AGROECOSYSTEMS

The basic tenets of a sustainable agroecosystem are conservation of renewable resources, adaptation of the crop-animal combinations to the environment, and maintenance of a moderate but sustainable level of productivity. To emphasize long-term ecological sustainability rather than short-term productivity, the production system must do the following:

1. Reduce energy and resource uses and regulate the overall energy input so that the output:input ratio is high
2. Reduce nutrient losses by effectively containing leaching, runoff, and erosion and improve nutrient recycling through the use of legumes, organic manure, compost, and other effective recycling mechanisms
3. Encourage local production of crops adapted to the natural and socioeconomic setting
4. Sustain a desired net output by preserving the natural resource base that exists by minimizing soil degradation or genetic erosion
5. Reduce costs and increase the efficiency and economic viability of small and medium-sized farms, thereby promoting a diverse, potentially resilient agricultural system (Altieri, 1987)

As shown in Figure 9.2, from a management viewpoint, the basic components of a sustainable agroecosystem include

1. Vegetative cover as an effective soil- and water-conserving measure, met through the use of no-till practices, mulch farming, use of cover crops, etc.
2. Regular supply of organic matter through regular addition of organic matter (manure, compost) and promotion of soil biotic activity
3. Nutrient-recycling mechanisms through the use of crop rotations, crop-livestock mixed systems, agroforestry, and intercropping systems based on legumes, etc.
4. Pest regulation assured through enhanced activity of biological control agents achieved through biodiversity manipulations and by introducing and/or conserving natural enemies

DESIGNING HEALTHY AGROECOSYSTEMS

As shown in the previous chapters, diversified cropping systems, such as those based on intercropping and agroforestry or cover cropping of orchards, have recently been the target of much research. This

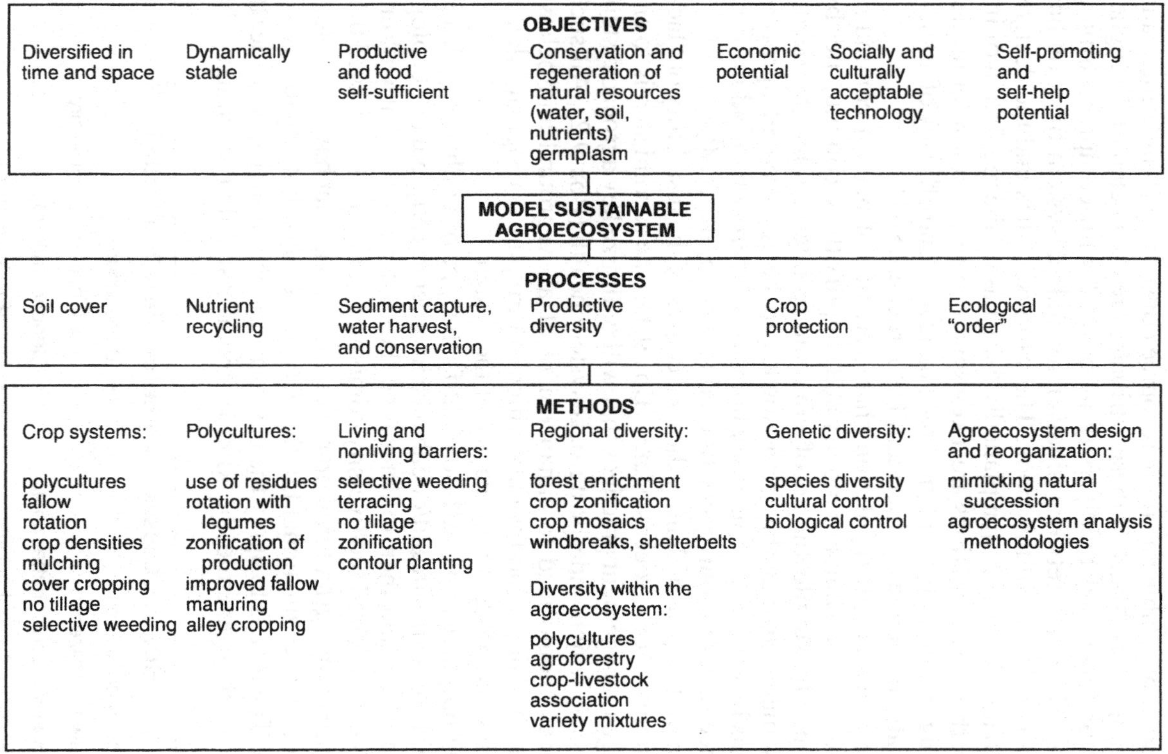

OBJECTIVES

- Diversified in time and space
- Dynamically stable
- Productive and food self-sufficient
- Conservation and regeneration of natural resources (water, soil, nutrients) germplasm
- Economic potential
- Socially and culturally acceptable technology
- Self-promoting and self-help potential

MODEL SUSTAINABLE AGROECOSYSTEM

PROCESSES

- Soil cover
- Nutrient recycling
- Sediment capture, water harvest, and conservation
- Productive diversity
- Crop protection
- Ecological "order"

METHODS

- Crop systems:
  - polycultures
  - fallow
  - rotation
  - crop densities
  - mulching
  - cover cropping
  - no tillage
  - selective weeding

- Polycultures:
  - use of residues
  - rotation with legumes
  - zonification of production
  - improved fallow
  - manuring
  - alley cropping

- Living and nonliving barriers:
  - selective weeding
  - terracing
  - no tillage
  - zonification
  - contour planting

- Regional diversity:
  - forest enrichment
  - crop zonification
  - crop mosaics
  - windbreaks, shelterbelts

  Diversity within the agroecosystem:
  - polycultures
  - agroforestry
  - crop-livestock association
  - variety mixtures

- Genetic diversity:
  - species diversity
  - cultural control
  - biological control

- Agroecosystem design and reorganization:
  - mimicking natural succession
  - agroecosystem analysis methodologies

FIGURE 9.2. Methods to diversify agroecosystems to ensure a series of ecological processes and sustainability objectives.

177

interest is largely based on the new emerging evidence that these systems are more stable and more resource conserving (Vandermeer and Perfecto, 1995). Many of these attributes are connected to the higher levels of functional biodiversity associated with complex farming systems. The link between biodiversity and ecosystem function has been a main focus of this book. The various studies and the designs used to test the effects that plant diversity has on the regulation of insect herbivore populations are a key source of information for implementing strategies that enhance the abundance and efficacy of associated natural enemies (Altieri and Letourneau, 1984).

The inherent self-regulation characteristics of natural communities are lost when humans modify and simplify these communities by breaking the fragile thread of community interactions. Agroecologists maintain that this breakdown can be repaired in agroecosystems by restoring community homeostasis through the addition or enhancement of biodiversity (Altieri and Nicholls, 2000).

The first step is to identify the root causes of the instability or "lack of immunity" of agroecosystems (Box 9.1). The second step is to encourage management practices that will optimize key agroecosystem processes, which underlie agroecosystem health (Box 9.2). All such practices should lead to enhancement of aboveground and belowground functional biodiversity which in turn play ecological roles in restoring the productive capacity of the system.

An important step is to also identify the type of biodiversity that is desirable to maintain and/or enhance in order to carry out key ecological services, and then to determine the best practices that will encourage the desired biodiversity components. Figure 1.5 showed that there are many agricultural practices and designs that have the potential to enhance functional biodiversity and others that negatively affect it. The idea is to apply the best management practices in order to enhance or regenerate the kind of biodiversity that subsidizes the

---

BOX 9.1. Causes of Agroecosystem Dysfunction

- Excess pesticides
- Excess fertilizers
- Low soil organic matter content
- Low soil biological activity

- Monoculture
- Low functional biodiversity
- Genetic uniformity
- Nutrient deficiencies
- Moisture imbalances

**BOX 9.2. Routes to Agroecosystem Health and Mechanisms
to Improve Agroecosystem Immunity**

Routes to agroecosystem health

- Strengthen the immune system (proper functioning of natural pest control)
- Decrease toxicity through elimination of agrochemicals
- Optimize metabolic function (organic matter decomposition and nutrient cycling)
- Balance regulatory systems (nutrient cycles, water balance, energy flow, population regulation, etc.)
- Enhance conservation and regeneration of soil-water resources and biodiversity
- Increase and sustain long-term productivity

Mechanisms to improve agroecosystem immunity

- Increase plant species and genetic diversity in time and space
- Enhance functional biodiversity (natural enemies, antagonists, etc.)
- Enhance soil organic matter and biological activity
- Increase soil cover and crop competitive ability
- Eliminate toxic inputs

---

health and sustainability of agroecosystems by providing ecological services such as biological pest control, nutrient cycling, water and soil conservation, etc. (Gliessman, 1999). As depicted in Figure 9.3, crop health can be achieved by establishing mechanisms that aid in the regulation of insect pests through two routes: (1) enhancing the rich natural enemy biodiversity harbored by a diversified agroecosystem, which has been explored in this book, and (2) encouraging a healthy soil rich in organic matter and a diverse soil biota.

HEALTHY SOILS—HEALTHY PLANTS

The fields of integrated pest management (IPM) and integrated soil-fertility management (ISFM) have proceeded separately without realizing that low-input agroecosystems rely on synergies of plant diversity and the continuing function of the soil microbial community and its relationship with organic matter to maintain the integrity of the agroecosystem. It is crucial for scientists to understand that most

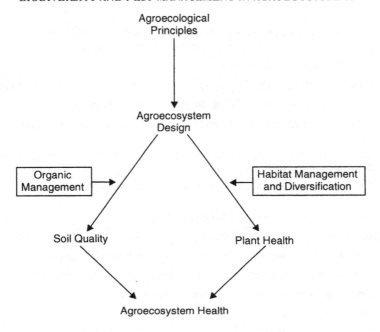

FIGURE 9.3. The pillars of egroecosystem health.

pest-management methods used by farmers can also be considered soil-fertility management strategies and that there are positive interactions between soils and pests that, once identified, can provide guidelines for optimizing total agroecosystem function (Figure 9.4). Increasingly, research is showing that the ability of a crop plant to resist or tolerate insect pests and diseases is tied to optimal physical, chemical, and especially biological properties of soils. Soils with high organic matter and active biological activity generally exhibit good soil fertility, as well as complex food webs and beneficial organisms that prevent infection. On the other hand, farming practices that cause nutrition imbalances can lower pest resistance (Phelan, Mason, and Stinner, 1995).

Yet it is for this reason that the crop-diversification strategies emphasized in this book must be complemented by regular applications of organic amendments (crop residues, animal manures, and composts) to maintain or improve soil quality and productivity. Much is

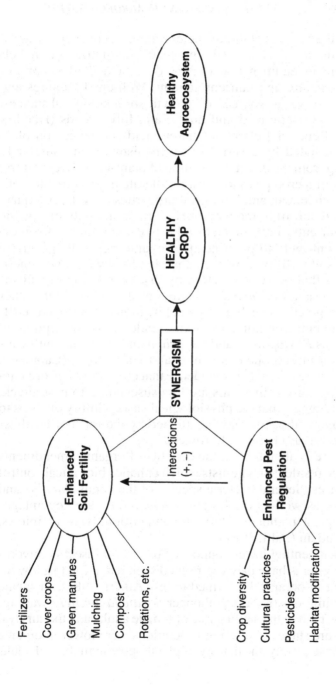

FIGURE 9.4. Interactions of soil and pest management practices used by farmers, some of which may result in synergism leading to a healthy and productive crop.

known about the benefits of multispecies rotations, cover crops, agro-forestry, and intercrops. Less well known are the multifunctional effects of organic amendments beyond the documented effects on improved soil structure and nutrient content. Well-aged manures and composts can serve as sources of growth-stimulating substances, such as indole-3 acetic acid and humic and fulvic acids (Hendrix et al., 1990). Beneficial effects of humic acid substances on plant growth are mediated by a series of mechanisms, many similar to those resulting from the direct application of plant-growth regulators.

The ability of a crop plant to resist or tolerate pests is tied to optimal physical, chemical, and biological properties of soils. Adequate moisture, good soil tilth, moderate pH, correct amounts of organic matter and nutrients, and a diverse and active community of soil organisms all contribute to plant health. Organic-rich soils generally exhibit good soil fertility as well as complex food webs and beneficial organisms that prevent infection by disease causing organisms such as *Phthium* and *Rhizoctonia* (Hendrix et al., 1990). On the other hand, farming practices such as high applications of nitrogen fertilizer can create nutrition imbalances and render crops susceptible to diseases such as *Phytophtora* and *Fusarium* and stimulate outbreaks of Homopteran insects such as aphids and leafhoppers (Campbell, 1989). In fact, there is increasing evidence that crops grown in organic-rich and biologically active soils are less susceptible to pest attack. Many studies suggest that the physiological susceptibility of crops to insect pests and pathogens may be affected by the form of fertilizer used (organic versus chemical fertilizer).

The literature is abundant on the benefits of organic amendments that encourage resident antagonists which enhance biological control of plant diseases. Several bacteria species of the genus *Bacillus* and *Pseudomonas,* as well as the fungus *Trichoderma,* are key antagonists that suppress pathogens through competition, lysis, antibiosis, or hyperparasitism (Palti, 1981).

Studies documenting lower abundance of several insect herbivores in low-input systems have partly attributed such reduction to a low nitrogen content in organically farmed crops (Altieri, 1985). In Japan, density of immigrants of the planthopper *Sogatella furcifera* was significantly lower while the settling rate of female adults and the survival rate of immature stages of ensuing generations were lower in organic rice fields. Consequently, the density of planthopper nymphs and adults in the ensuing generations decreased in organically farmed fields (Kajimura, 1995). In England, conventional winter wheat fields developed a larger infestation of the aphid *Metopolophium dirhodum* than its organic counterpart. This crop also had higher levels of free protein amino acids in its leaves during June, which were believed to have resulted from a nitrogen top dressing of the crop early in April. However, the difference in the aphid infestation between crops was attributed to the aphid's response to relative proportions of certain nonprotein to protein amino acids in the leaves at the time of aphid settling on crops (Kowalski and Visser, 1979). In greenhouse experiments, when given a choice of maize grown on organic versus chemically treated soils, European corn borer females preferred to lay significantly more eggs in chemically fertilized plants (Phelan, Mason, and Stinner, 1995).

RESTORING DIVERSITY IN AGRICULTURAL SYSTEMS

Most regions in the world have many types of agricultural systems determined by local variations in climate, soil, economic relations, social structure, and history. Clearly, these systems are always changing in size, land-tenure assignments, technological intensity, population shifts, resource availability, environmental degradation, economic growth or stagnation, political change, and so forth. Farmers adapt to some of these changes by responding through technological innovation to variations in the physical environment, prices of inputs, etc. A logical outcome to present environmental and socioeconomic constraints is the desire by farmers for more sustainable agricultural methods.

The basic concepts of a self-sustaining, low-input, diversified, and efficient agricultural system must be synthesized into practical alternative systems to suit the specific needs of farming communities in different agroecological regions of the world. A major strategy in sustainable agriculture is to restore agricultural diversity in time and space through crop rotations, cover crops, intercropping, crop-livestock mixtures, etc. (Altieri, 1987). As seen in Figure 9.5, different options to diversify cropping systems are available depending on whether the current monoculture systems to be modified are based on annual or perennial crops. Diversification can also take place outside of the farm, for example, in crop-field boundaries with windbreaks,

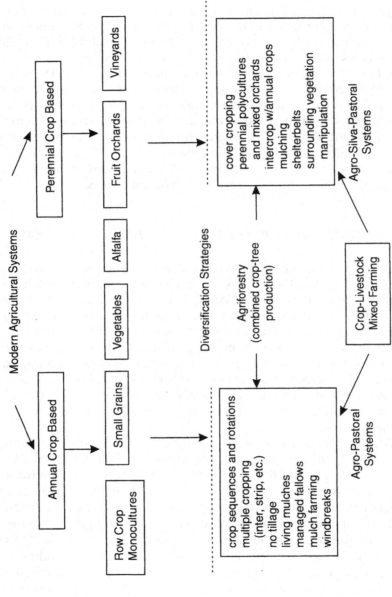

FIGURE 9.5. Diversification strategies for annual crop-based or perennial crop-based modern agroecosystems.

shelterbelts, and living fences or corridors, which can improve habitat for wildlife and beneficial insects, provide sources of wood, organic matter, and resources for pollinating bees, and, in addition, modify wind speed and the microclimate (Altieri and Letourneau, 1982; Kemp and Barrett, 1989).

In this book many alternative diversification strategies that exhibit beneficial effects on soil fertility, crop protection, and crop yields have been explored. Key ones include the following:

1. *Crop Rotations:* Temporal diversity incorporated into cropping systems, providing crop nutrients, and breaking the life cycles of several insect pests, diseases, and weeds
2. *Polycultures:* Complex cropping systems in which two or more crop species are planted within sufficient spatial proximity to result in competition or complementation, thus enhancing yields (Francis, 1986; Vandermeer, 1989)
3. *Agroforestry Systems:* An agricultural system where trees are grown together with annual crops and/or animals, resulting in enhanced complementary relations between components, increasing multiple use of the agroecosystem (Nair, 1993)
4. *Cover Crops:* The use of pure or mixed stands of legumes or other annual plant species under fruit trees for the purpose of improving soil fertility, enhancing biological control of pests, and modifying the orchard microclimate (Finch and Sharp, 1976)
5. *Animal Integration:* In agroecosystems this aids in achieving high biomass output and optimal recycling

If these alternative technologies are used, the possibilities of complementary interactions between agroecosystem components are enhanced, resulting in one or more of the following effects:

1. Continuous vegetation cover for soil protection
2. Constant production of food, ensuring a varied diet and several marketing items
3. Closing of nutrient cycles and effective use of local resources
4. Soil and water conservation through mulching and wind protection

5. Enhanced biological pest control through diversification which provides resources to beneficial biota
6. Increased multiple-use capacity of the landscape and/or sustained crop production without relying on environmentally degrading chemical inputs

In summary, key ecological principles for design of diversified and sustainable agroecosystems include the following:

1. *Increasing species diversity,* as this promotes fuller use of resources (nutrients, radiation, water, etc.), pest protection, and compensatory growth. Many researchers have highlighted the importance of various spatial and temporal plant combinations to facilitate complementary resource use or to provide intercrop advantage such as in the case of legumes facilitating the growth of cereals by supplying extra nitrogen. Compensatory growth is another desirable trait: as one species succumbs to pests, weather, or harvest, another species fills the void and maintains full use of available resources. Crop mixtures also minimize risks by creating the sort of vegetative texture that controls specialist pests.
2. *Enhance longevity* through the addition of perennials that contain a thick canopy, thus providing continual cover that can also protect the soil. Constant leaf fill builds organic matter and allows uninterrupted nutrient circulation. Dense, deep root systems of long-lived woody plants are an effective mechanism for nutrient capture, offsetting the negative losses through leaching.
3. *Impose a fallow* to restore soil fertility through biologically mediated mechanisms and to reduce agricultural pest populations as life cycles are interrupted with forest regrowth.
4. *Enhance additions* of organic matter by including high biomass-producing plants. Accumulation of both "active" and "slow fraction" organic matter is key for activating soil biology, improving soil structure and macroporosity, and elevating the nutrient status of soils.
5. *Increase landscape diversity* by having in place a mosaic of agroecosystems representative of various stages of succession. Risk of complete failure is spread among, as well as within, the various cropping systems. Improved pest control is also linked to spatial heterogeneity at the landscape level.

ENHANCING SURROUNDING BIODIVERSITY

As discussed in Chapter 7, manipulation of vegetation adjacent to agricultural fields can be key in providing overwintering sites and alternative food sources for entomophagous arthropods. The impact of such border vegetation or refuges is dependent on its plant composition and the spatial extent of its influence on natural enemy abundance, which is determined by the distance to which natural enemies. disperse into the crop (Corbett and Rosenheim, 1996). Other authors have emphasized the importance of landscape complexity, such as large fallows, riparian forests, or other features near crop fields. Research has shown that natural enemy abundance and efficiency increases with landscape heterogeneity while pest damage increases as the percentage of noncrop area in the landscape decreases (Thies and Tscharntke, 1999). Such observations suggest that in order to maximize the impact of natural enemies through habitat management, · agroecologists must look beyond the immediate confines of agricultural lands to include the uncultivated habitats that separate or surround cultivated fields. Landis and Marino (1996a) argue that in order to effectively conserve natural enemies in early-successional agricultural landscapes the creation and management of mid- to late-successional habitats may be required. In essence, this is a process of refragmenting highly disturbed landscapes by adding a network of more stable habitats of varying successional stages. These habitats should serve multiple functions as cross-wind trap strips, filter strips, riparian buffer zones, or agroforestry production systems (Landis and Marino, 1996b).

There is sufficient information on certain forms of cultural and biological control applicable to some crop pests of known biology. Based on such information, Perrin (1980) advanced a series of environmental management proposals to improve the control of insect pests affecting the cereal-rape system in southern England. Although Perrin suggests some important changes in the design of cereal-rape systems, the protocol does not address some important dilemmas, such as whether it is desirable to have hedgerows removed or stop aerial spraying of insecticides. Nevertheless, Perrin's proposal is a step in the right direction in that it advances a regional approach where landscape diversity is manipulated in a coordinated manner by all agricultural sectors involved. The possibilities of such cooperation are not encouraging when antagonistic production and conservation views prevail. Such conflicts are well illustrated by the debate on hedges in England, where on the one hand their removal increases the efficiency with which the land can be farmed with modern machinery, but on the other may decrease the local diversity and abundance of birds, insects, and plants. In these cases, an agroecological approach must be developed so that economic, social, and environmental goals are defined by the local rural community, and low-input technologies are implemented to harmonize economic growth, social equity, and environmental preservation (Figure 9.6).

CASE STUDY 1: DIVERSIFICATION OF AN ONION AGROECOSYSTEM IN MICHIGAN

In order to optimize the mortality of the major onion insect pest (onion maggot) in Michigan, a functionally diverse onion agroeco-

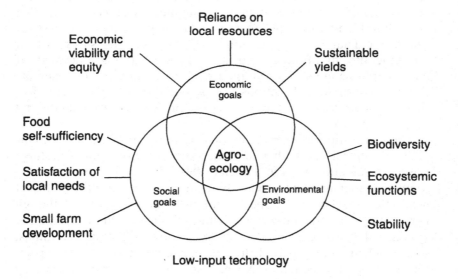

FIGURE 9.6. The role of agroecology in satisfying social, environmental, and economic goals in rural areas (after Altieri, 1995).

system was designed. This design was derived from quantitative models describing the relationships among components in the system. From an understanding of these quantitative interactions, designs incorporating diseases, weeds, insects, etc., can be derived as long as the relationships that are used in the construction of these "free-body" models are structure independent or incorporate aspects of structure as a variable.

The alternative design of the onion agroecosystem shown in Figure 9.7 stresses planned or functional diversity. The cow pasture and weedy borders provide alternate host and nectar for the onion maggot parasite *Aphaereta pallipes* (Groden, 1982). The cow pasture also provides a rich resource for earthworms, thereby potentially maximizing the densities of the tiger fly predator of onion flies. The long, narrow strips of onions minimize the distance from any point in the onion field and the weedy borders and cow pasture. This is important since *A. pallipes* numbers decline exponentially from weedy borders and cow pastures into the onion field (Groden, 1982). This is also true of onion flies infected with disease caused by *Entomophthora muscae.* Weedy field borders are not mowed so that attachment sites for diseased flies are provided. Narrow weedy borders maximize the probability of *E. muscae* spores encountering healthy flies by crowding together resting and attachment sites for healthy flies during midday. By mowing some of the weedy border, this crowding effect can be increased. The planting of radishes adjacent to onions provides an alternate host and thus a continuous food supply for the rove beetle *Aleochara bilineata.* A number of plantings should be used in order to provide a season-long food resource for the cabbage maggot, and a number of different planting dates of onions should be incorporated into the design (Groden, 1982). Groden also showed that early-planted onions adjacent to late-planted onions serve as a highly attractive trap crop resulting in a concentration of the onion maggot population in the early planting. Because the later plantings go largely untouched, the early planting can be positioned near the radish interface so that the hosts pool for *A. bilineata* is concentrated, thereby making prey search more efficient.

In order to deal with the problem of emerging flies after onions are harvested, management of cull onions becomes a major issue. A diversification-management option involves sowing a fall rye or oat cover crop immediately after harvest so that in a week the cover crop

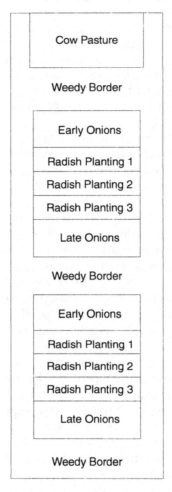

FIGURE 9.7. Sustainable agriculture planting for minimizing the impact of onion maggot and the need for the use of insecticides for control of this pest (after Groden, 1982).

hides the cull onions in the field, making it difficult for the onion flies to find the culls. A modification is not to harvest a small section of onion rows, and then, while sowing the cover crop, the tops of the onions can be cut off and left on the ground. These cut tops are very attractive to the onion flies (more attractive than cull onions); however, the immature onion fly cannot survive on them because they dry up before insect development can be completed. Thus, the cut tops serve to keep the onion flies from laying on the culls until the cover crop comes up, and then the searching efficiency of the female flies is drastically reduced. In addition, crop rotation significantly reduces the number of flies colonizing an onion field in the spring (Mortinson, Nyrop, and Eckenroad, et al., 1988).

CASE STUDY 2: A DIVERSIFIED
SMALL FARMING SYSTEM IN CHILE

Since 1980, the Centro de Educacion y Tecnologia (CET), a Chilean nongovernmental organization, has engaged in a rural development program aimed at helping resource-poor peasants to achieve year-round food self-sufficiency and rebuild the productive capacities of their small land holdings. The CET's approach has been to establish several 0.5 ha model farms where most of the food requirements for a family with scarce capital and land can be met. In this farm system, the critical factor in the efficient use of scarce resources is diversity.

The CET farm is a diversified combination of crops, trees, and animals. In an attempt to maximize production efficiencies, the components are structured to minimize losses to the system and promote positive interactions. Thus, crops, animals, and other farm resources are managed to optimize production efficiency, nutrient and organic matter recycling, and crop protection. The principal components include the following:

- Vegetables: spinach, cabbage, tomatoes, lettuce, etc.
- Intercropped maize-beans-potatoes and peas-fava beans
- Cereals: wheat, oats, and barley
- Forage crops: clover, alfalfa, and ryegrass
- Fruit trees: grapes, oranges, peaches, apples, etc.
- Nonfruit trees: black locust, honey locust, willows, etc.
- Domestic animals: chickens, pigs, ducks, goats, bees, and dairy cow

The animal and plant components are chosen according to (1) the crop or animal nutritional contributions, (2) their adaptation to local agroclimatic conditions, (3) local peasant consumption patterns, and (4) market opportunities. The design is also based on cropping patterns, crops, and management techniques practiced by local campesinos (small farmers). In Chile, campesinos typically produce a great variety of crops and animals. It is not unusual to find as many as five or ten tree crops, ten to fifteen annual crops, and three to five animal species on a single farm.

The physical layout of these model farms varies depending on local conditions; however, most vegetables are produced in heavily composted raised beds located in the garden section, each of which can produce up to 83 kg of fresh vegetables per month. The rest of the vegetables, cereals, legumes, and forage plants are produced in a six-year rotational system (Figure 9.8). This rotation was designed to provide the maximum variety of basic crops in six plots, taking advantage of the soil-restoring properties of rotations. Each plot received the following treatments during the six-year period (Figure 9.9):

Year 1: Summer: corn, beans, and potato
         Winter: peas and fava beans
Year 2: Summer: tomato, onion, and squash
         Winter: supplementary pasture (oats, clover, ryegrass)
Year 3: Summer: soybean, peanuts, or sunflower
         Winter: wheat companion—planted with pasture
Year 4-6: Permanent pasture: clover, alfalfa, and ryegrass

In each plot, crops are grown in several temporal and spatial designs, such as strip cropping, intercropping, mixed cropping, cover crops, and living mulches, optimizing the use of limited resources and enhancing the self-sustaining and resource-conserving attributes of the system. An important consideration in the rotational design was the stability of the cropping system in terms of soil-fertility maintenance and pest regulation. It is well accepted that a rotation of grains and leguminous forage crops provides significant additional inputs of nitrogen and much higher yields of the subsequent crop of grain obtained with continuous grain monocropping. The output of grain depends on the efficiency of the legumes in supplying nitrogen. Studies in Chile, and elsewhere, have shown that legumes such as sweet clover, alfalfa, and hairy vetch can produce between 2.3 and 10 tons/ha of dry matter and fix between 76 and 367 kg of nitrogen/ha.

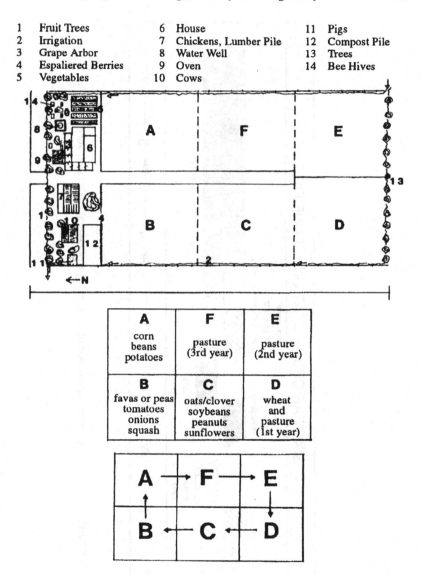

1	Fruit Trees	6	House	11	Pigs
2	Irrigation	7	Chickens, Lumber Pile	12	Compost Pile
3	Grape Arbor	8	Water Well	13	Trees
4	Espaliered Berries	9	Oven	14	Bee Hives
5	Vegetables	10	Cows		

A	F	E
corn beans potatoes	pasture (3rd year)	pasture (2nd year)
B	**C**	**D**
favas or peas tomatoes onions squash	oats/clover soybeans peanuts sunflowers	wheat and pasture (1st year)

FIGURE 9.8. Model design of a self-sufficient farming system, based on a six-year rotational scheme adaptable to Mediterranean environments (adapted from Altieri, 1987).

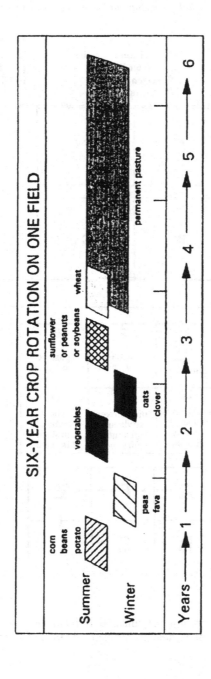

FIGURE 9.9. Crop sequence in a six-year rotation design for central Chile (after Altieri, 1987).

This is sufficient to meet most of the nitrogen requirements of agronomic and vegetable crops.

The rotational scheme provides nearly continuous plant cover that aids in the control of annual weeds. Incorporating legume cover crops in annual crops, such as corn, cabbage, and tomato, through over-seeding and sod-based rotations, has reduced weeds significantly. In addition, these systems reduce erosion, conserve moisture, and, therefore, offer great potential for hillside farmers.

The crop rotations scheme also had a profound impact on insect pest populations. For example, the corn rootworm (*Diabrotica* spp.) consistently reached higher levels in a continuous corn monoculture than in cornfields following soybean, clover, alfalfa, or other crops. The presence of alfalfa in the rotational scheme enhanced the abundance and diversity of insect predators and parasites on the farm. Strip cutting of alfalfa forced movement of predators from alfalfa to other crops. Cutting and spreading alfalfa hay, containing high numbers of beneficial insects, throughout the farm also increased natural enemy populations. Cereal residues used as straw mulches in the succeeding crops significantly reduced whitefly populations, the principal vector of several viruses, by affecting their attraction to host crops. Spiders, ground beetles, and other predators also were enhanced by the alternative habitat provided by the mulch.

CET personnel have closely monitored the performance of this integrated farming system. Throughout the years, soil fertility has improved (P_2O_5 levels, which were initially declining, increased from 5 to 15 ppm) and no serious pest or disease problems have been noticed. The fruit trees in the garden orchard and those surrounding the rotational plots produce about 843 kg of fruit/year (grapes, quince, pears, plums, etc.). Forage production reaches about 18 tons/0.21 ha per year, milk production averages 3,200 L per year, and egg production reaches a level of 2,531 units. A nutritional analysis of the system based on the production of the various plant and animal components (milk, eggs, meat, fruits, vegetables, honey, etc.) shows that the system produces a 250 percent surplus of protein, 80 percent and 550 percent surplus of vitamin A and C respectively, and a 330 percent surplus of calcium. A household economic analysis indicates that given a list of preferences, the balance between selling surpluses and buying preferred items is a net income of US$790. If all the farm output is sold at wholesale prices, the family could generate a net income of US$1,635, equivalent to a monthly income of US$136, 1.5 times greater than the monthly legal minimum wage in Chile (Yurjevic, 1991).

Conclusion

Typical commercial-production agriculture has resulted in the simplification of cropping systems in general. The expansion of monocultures has decreased abundance and activity of natural enemies due to the removal of critical food resources and overwintering sites (Corbett and Rosenheim, 1996). Many scientists are concerned that with accelerating rates of habitat removal, the contribution to pest suppression by biocontrol agents using these habitats will decline (Fry, 1995; Sotherton, 1984). For this reason, many researchers cited in this book have proposed options at the field and landscape level to rectify this decline by increasing the vegetational diversity of agroecosystems and surrounding areas.

This book has examined hundreds of studies which show that complementary interactions occur between crops and/or between crops and weeds grown in polycultures and between adjacent cultivated and uncultivated vegetational components of agroecosystems. These interactions can have positive or negative, direct or indirect effects on the biological control of specific crop pests. The exploitation of these interactions in real situations involves agroecosystem design and management and requires an understanding of the numerous relationships among plants, herbivores, and natural enemies (Altieri and Letourneau, 1982). Clearly, the emphasis of this approach is to restore natural control mechanisms through the addition of selective diversity, rather than forcing the establishment of biological control in simplified environments (such as monocultures) where the essential ecological elements are lacking to allow for optimum performance of natural enemies (Van Driesche and Bellows, 1996).

A key strategy in agroecology is to exploit the complementarity and synergy that result from the various combinations of crops, trees, and animals in agroecosystems that feature spatial and temporal arrangements such as polycultures, agroforestry systems, and crop-livestock mixtures. This implies identifying the type of biodiversity that is desirable to maintain and/or enhance in order to carry out ecological services, and then to determine the best practices that will encourage the desired biodiversity components. Many agricultural practices and designs have the potential to enhance functional biodiversity and others negatively affect it. The idea is to apply the best management practices in order to enhance or regenerate the kind of biodiversity that can best subsidize the sustainability of agroecosystems by providing ecological services such as biological pest control, nutrient cycling, water and soil conservation, and so forth. The role of agroecologists should be to encourage those agricultural practices that increase the abundance and diversity of aboveground and belowground organisms, which in turn provide key ecological services to agroecosystems (Altieri and Nicholls, 2000).

In order for this diversification strategy to be more rapidly implemented, it is necessary to develop a much better understanding of the ecology of parasitoids and predators within and outside of the cultivated habitat, identifying those resources that are necessary for their survivorship and reproduction (Gurr, Van Emden, and Wratten, 1998). It is also important to determine to what extent populations within the crop contribute to the overall natural enemy metapopulation in subsequent years. If these contributions are minor, then investments in habitat management should be oriented specifically to increasing the source populations outside the crop to ensure a greater number of immigrants each year, an action parallel to increasing the dosage of a chemical biocide. However, if the subpopulations within the cropping system contribute significantly to the year-to-year metapopulation dynamics, then habitat modifications should consider not only tactics fostering immigration into the crop but also those augmenting the probability of successional emigration when this habitat becomes unsuitable. Such actions could include the addition of plant species to provide alternate hosts and/or food sources, the addition of habitats as suitable overwintering sites or the provision of corridors within the cropping system to facilitate movement between the subcomponents of the metapopulation.

In summary, there are four key issues to consider when implementing habitat management:

1. The selection of the most appropriate plant species and their spatial/temporal deployment
2. The predator-parasitoid behavioral mechanisms which are influenced by the manipulation

3. The spatial scale over which the habitat enhancement operates
4. The potential negative aspects associated with adding new plants to the agroecosystem (Landis, Wratten, and Gurr, 2000)

Obviously, proposed habitat-management techniques must fit existing cropping systems and adapt to the needs and circumstances of farmers.

Prokopy (1994) has cautioned that all diversification interactions should be evaluated within the context of a broader integrated management program of the agricultural crop. The reason for caution is that potential benefits may be less than unforeseen costs. For example, although blackberry plants in California serve as a host plant for alternative hosts of the parasitoid *A. epos* these same plants could be a reservoir for the bacterium responsible for Pierce's disease, a serious disease of grapes transmitted by the blue-green sharpshooter. Thus, an action taken to increase the efficacy of natural enemies could incur losses through increased levels of disease. However, actions to reduce Pierce's disease, such as removal of sharpshooter host plants within riparian habitats as recommended in California's Napa Valley, can in turn lead to a reduction of natural enemies of grape leafhoppers, compounding a minor pest problem. Clearly then, it is important to provide the right kind of diversity. At times the addition of one or two plants is all it may take.

Recent studies conducted in grassland systems suggest that there are no simple links between species diversity and ecosystemic stability. What is apparent is that functional characteristics of component species are as important as the total number of species. Recent experiments with grassland plots conclude that functionally different roles represented by plants are at least as important as the total number of species in determining processes and services in ecosystems (Tilman, Wedin, and Knops, 1996).

This latest finding has practical implications for agroecosystem management. If it is easier to mimic specific ecosystem processes rather than duplicate all the complexity of nature, then the focus should be placed on incorporating a specific biodiversity component that plays a specific role, such as a plant that fixes nitrogen, provides cover for soil protection, or harbors resources for natural enemies. In the case of farmers without major economic and resource limitations who can afford a certain risk of crop failure, a crop rotation or a simple crop association may be all it takes to achieve a desired level of stability. In the case of resource-poor farmers, where crop failure is intolerable, highly diverse polyculture systems would probably be the best choice. The obvious reason is that the benefit of complex agroecosystems is low risk; if a species falls to disease, pest attack, or weather, another species is available to fill the void and maintain a full use of resources.

The central issue in sustainable agriculture is not achieving maximum yield—it is long-term stabilization (Reganold et al., 2001). Sustaining agricultural productivity will require more than a simple modification of traditional ad hoc techniques. The development of self-sufficient, diversified, economically viable agroecosystems comes from novel designs of cropping and/or livestock systems managed with technologies adapted to the local environment that are within the farmers' resources. Energy and resource conservation, environmental quality, public health, and equitable socioeconomic development should be considered in making decisions on crop species, rotations, row spacing, fertilizing, pest control, and harvesting. Many farmers will not shift to alternative systems unless there is a good prospect for monetary gain, brought about by either increased output or decreased production costs. Different attitudes will depend primarily on farmers' perceptions of the short-term and long-term economic benefits of sustainable agriculture.

Restoration of natural controls in agroecosystems through vegetation management not only regulates pests but also helps to conserve energy, improve soil fertility, minimize risks, and reduce dependence on external resources. The ultimate goal of agroecological design is to integrate components so that overall biological efficiency is improved, biodiversity is preserved, and the agroecosystem productivity and its self-sustaining capacity is maintained. The goal is to design a quilt of agroecosystems within a landscape unit, each mimicking the structure and function of natural ecosystems, that is, systems that include the following:

1. Vegetative cover as an effective soil- and water-conserving measure, met through the use of no-till practices, mulch farming, and use of cover crops and other appropriate methods
2. A regular supply of organic matter through the regular addition of organic matter (manure, plant biomass, compost, and promotion of soil biotic activity)

3. Nutrient-recycling mechanisms through the use of crop rotations, crop-livestock systems based on legumes, etc.
4. Pest regulation assured through enhanced activity of biological control agents achieved by introducing and/or conserving through vegetational designs, natural enemies, and antagonists

This is particularly important in underdeveloped countries where sophisticated inputs are either not available or may not be economically or environmentally advisable, especially to resource-poor farmers. Further research in this area should provide an ecological basis for the design of diverse, pest-stable, self-sustained agroecosystems. These systems are urgently needed worldwide in an era of deteriorating environmental quality, a worsened energy situation, and escalating input costs. This approach to agriculture will be practical only if it is economically sensible and can be carried out within the constraints of a fairly normal agricultural-management system. However, given the trend toward large-scale, specialized farm production throughout the world, objectively there is not much room left for a fair implementation of a regional insect-habitat management program. Emerging biotechnological approaches such as transgenic crops deployed in more than 40 million hectares in 2000 are leading agriculture toward further specialization, and the potential effects of transgenic crops on nontarget beneficial organisms is of concern to biological control practitioners (Rissler and Mellon, 1996; Hilbeck et al., 1998; Altieri, 2000).

Long-term maintenance of diversity requires a management strategy that considers regional biogeography and landscape patterns, as well as design of environmentally sound agroecosystems above purely economic concerns. This is why several authors have repeatedly questioned whether the pest problems of modern agriculture can be ecologically alleviated within the context of the present capital-intensive structure of agriculture. Buttel (1980) suggests that many problems of modern agriculture are rooted within that structure and calls for the consideration of major social change, land reform, redesign of machinery, research, and extension reorientation in the agricultural sector to increase the possibilities of improved pest control through vegetation management. Whether the potential and spread of ecologically based pest management is realized will depend on policies, attitude changes on the part of researchers and policymakers, existence of markets for organic produce, and also farmer and consumer movements that demand a more healthy and viable agriculture.

It is crucial that scientists involved in the search for sustainable agricultural technologies be concerned about who will ultimately benefit from them. Thus, what is produced, how it is produced, and for whom it is produced are key questions that need to be addressed if a socially equitable agriculture is to emerge. When such questions are examined, issues of land tenure, labor, appropriate technology, public health, and research policy unavoidably arise. Undoubtedly these are key challenges to deal with when developing a sustainable agriculture in the twenty-first century. Such issues must be urgently addressed by committed scientists and farmers in a true partnership.

References

Adams, J. B. and M. E. Drew. 1965. Grain aphids in New Brunswick. II. Aphid populations in herbicide-treated oatfields. *Canadian Journal of Zoology* 43: 789-794.

Adams, M. W., A. H. Ellingbae, and E. C. Rossineau. 1971. Biological uniformity and disease epidemics. *BioScience* 21: 1067-1070.

Aldrich, R. J. 1984. *Weed-crop ecology—Principles in weed management.* Breton Publishers, North Scituate, MA.

Alghali, A. M. 1993. Intercropping as a component in insect pest management for grain cowpea production in Nigeria. *Insect Science and Applications* 14: 49-54.

Ali, A. D. and T. E. Reagan. 1985. Vegetation manipulation impact on predator and prey populations in Louisiana sugar-cane ecosystems. *Journal of Economic Entomology* 7-8: 1409-1414.

Ali, M. I. and M. A. Karim. 1989. The use of trap crop in manipulating population of the cotton jassid on cotton. *Bangladesh Journal of Zoology* 17: 159-164.

Allen, W. W. and R. F. Smith. 1958. Some factors influencing the efficiency of *Apanteles medicagines* Muesebeck (Hymenoptera: Braconidae) as a parasite of the alfalfa caterpillar, *Colias philodice eurytheme* Boisduval. *Hilgardia* 28: 1.

Altieri, M. A. 1984. Patterns of insect diversity in monocultures and polycultures of brussels sprouts. *Protection Ecology* 6: 227-232.

Altieri, M. A. 1987. *Agroecology: The scientific basis of alternative agriculture.* Westview Press, Boulder, CO.

Altieri, M. A. 1991a. Ecology of tropical herbivores in polycultural agroecosystems. In *Plant-animal interactions: Evolutionary ecology in tropical and temperate regions,* pp. 607-617. P. W. Price, T. M. Lewinsohn, G. W. Fernandez, and W. W. Benton, eds. Wiley, New York.

Altieri, M. A. 1991b. How best can we use biodiversity in agroecosystems? *Outlook on Agriculture* 20: 15-23.

Altieri, M. A. 1991c. Increasing biodiversity to improve insect pest management in agro-ecosystems. In *Biodiversity of microorganisms and invertebrates: Its role in sustainable agriculture,* pp. 165-182. D. L. Hawksworth, ed. CAB Int., Wallingford, UK.

Altieri, M. A. 1991d. Traditional farming in Latin America. *The Ecologist* 21: 93-96.

Altieri, M. A. 1995. *Agroecology: The science of sustainable agriculture.* Westview Press, Boulder, CO.

Altieri, M. A. 1999. The ecological role of biodiversity in agroecosystems. *Agriculture, Ecosystems and Environment* 74: 19-31.

Altieri, M. A. 2000. The ecological impacts of transgenic crops on agroecosystem health. *Ecosystem Health* 6: 13-23.

Altieri, M. A. and J. D. Doll. 1978. Some limitations of weed biocontrol in tropical crop ecosystems in Colombia. In *Proceedings IV International Symposium on Biological Control of Weeds,* pp. 74-82. T. E. Freeman, ed. University of Florida, Gainesville, FL.

Altieri, M. A., C. A. Francis, A. Schoonhoven, and J. Doll. 1978. A review of insect prevalence in maize (*Zea mays* L.) and bean (*Phaseolus vulgaris* L.) polycultural systems. *Field Crops Research* 1: 33-49.

Altieri, M. A. and S. R. Gliessman. 1983. Effects of plant diversity on the density and herbivory of the flea beetle, *Phyllotreta cruciferae* Goeze, in California collard *(Brassica oleracea)* cropping systems. *Crop Protection* 2: 497-501.

Altieri, M. A. and S. B. Hecht. 1991. *Agroecology and small farm development.* CRC Press, Boca Raton, FL.

Altieri, M. A. and D. K. Letourneau. 1982. Vegetation management and biological control in agroecosystems. *Crop Protection* 1: 405-430.

Altieri, M. A. and D. K. Letourneau. 1984. Vegetation diversity and insect pest outbreaks. *CRC Critical Reviews in Plant Sciences* 2: 131-169.

Altieri, M. A., W. J. Lewis, D. A. Nordlund, R. C. Gueldner, and J. W. Todd. 1981. Chemical interactions between plants and *Trichogramma* wasps in Georgia soybean fields. *Protection Ecology* 3: 259-263.

Altieri, M. A. and M. Liebman. 1988. Weed management: Ecological guidelines. In *Weed management in agroecosystems: Ecological approaches,* pp. 183-218. M. A. Altieri and M. Z. Liebman, eds. CRC Press, Boca Raton, FL.

Altieri, M. A. and C. I. Nicholls. 2000. Applying agroecological concepts to development of ecologically based pest management systems. In *Proceedings of a Workshop "Professional societies and ecological based pest management systems,"* pp. 14-19. National Research Council, Washington, DC.

Altieri, M. A. and L. L. Schmidt. 1985. Cover crop manipulation in northern California orchards and vineyards: Effects on anthropod communities. *Biological Agriculture and Horticulture* 3: 1-24.

Altieri, M. A. and L. L. Schmidt. 1986a. The dynamics of colonizing arthropod communities at the interface of abandoned organic and commercial apple orchards and adjacent woodland habitats. *Agriculture, Ecosystems and Environment* 16: 29-43.

Altieri, M. A. and L. L. Schmidt. 1986b. Population trends and feeding preferences of flea beetles (*Phyllotreta cruciferae* Goeze) in collard-wild mustard mixtures. *Crop Protection* 5: 170-175.

Altieri, M. A. and L. L. Schmidt. 1987. Mixing cultivars of broccoli reduces cabbage aphid populations. *California Agriculture* 41: 24-26.

Altieri, M. A., A. V. Schoonhoven, and J. D. Doll. 1977. The ecological role of weeds in insect pest management systems: A review illustrated with bean (*Phaseolus vulgaris* L.) cropping systems. *PANS* 23: 195-205.

Altieri, M. A. and J. W. Todd. 1981. Some influences of vegetational diversity on insect communities of Georgia soybean fields. *Protection Ecology* 3: 333-338.

Altieri, M. A., J. W. Todd, E. W. Hauser, M. Patterson, G. A. Buchanan, and R. H. Walker. 1981. Some effects of weed management and row spacing on insect abundance in soybean fields. *Protection Ecology* 3: 339-343.

Altieri, M. A. and W. H. Whitcomb. 1979a. Manipulation of insect patterns through seasonal disturbance of weed communities. *Protection Ecology* 1: 185-202.

Altieri, M. A. and W. H. Whitcomb. 1979b. The potential use of weeds in the manipulation of beneficial insects. *Hort. Science* 14: 12-18,

Altieri, M. A. and W. H. Whitcomb. 1980. Weed manipulation for insect management in corn. *Environmental Management* 4: 483-489.

Altieri, M. A., R. C. Wilson, and L. L. Schmidt. 1985. The effects of living mulches and weed cover on the dynamics of foliage and soil arthropod communities in three crop systems. *Crop Protection* 4: 201-213.

Anderson, R. N. 1968. *Germination and establishment of weeds for experimental purposes.* Weed Science Society of America, USDA, Washington, DC.

Andow, D. 1983a. Effect of agricultural diversity on insect populations. In *Environmentally sound agriculture*, pp. 91-115. W. Lockeretz, ed. Praeger, New York.

Andow, D. A. 1983b. The extent of monoculture and its effects on insect pest populations with particular reference to wheat and cotton. *Agriculture, Ecosystems and Environment* 9: 25-35.

Andow, D. A. 1991a. Vegetational diversity and arthropod population response. *Annual Review of Entomology* 36: 561-586.

Andow, D. A. 1991b. Yield loss to arthropods in vegetational diverse agroecosystems. *Environmental Entomology* 20: 1228-1235.

Andow, D. A., A. G. Nicholson, H. C. Wien, and H. R. Wilson. 1986. Insect populations on cabbage grown with living mulches. *Environmental Entomology* 15: 293-299.

Andow, D. A. and D. R. Prokrym. 1990. Plant structural complexity and host finding by a parasitoid. *Oecologia* 62: 162-165.

Andow, D. and S. J. Risch. 1985. Predation in diversified agroecosystems: Relations between a coccinellid predator *Coleomegilla maculata* and its food. *Journal of Applied Ecology* 22: 357-372.

Andrews, D. J. and A. H. Kassam. 1976. The importance of multiple cropping in increasing world food supplies. In *Multiple cropping*, pp. 1-10. ASA Special Publication No. 27. G. B. Triplett, P. A. Sanchez and R. I. Papendick, eds. American Society of Agronomy, Madison, WI.

Aveling, C. 1981. The role of *Anthocoris* species (Hemiptera: Anthocoridae) in the integrated control of the damson-hop aphid *(Phorodon humuli)*. *Annals of Applied Biology* 97: 143.

Bach, C. E. 1980a. Effects of plant density and diversity on the population dynamics of a specialist herbivore, the striped cucumber beetle, *Acalymma vittata* (Fab.). *Ecology* 61: 1515-1530.

Bach, C. E. 1980b. Effects of plant diversity and time of colonization on an herbivore-plant interaction. *Oecologia* 44: 319-326.

Baliddawa, C. W. 1985. Plant species diversity and crop pest control: An analytical review. *Insect Science and Applications* 6: 479-487.

Bantilan, R. T., M. C. Palada, and R. R. Harwood. 1974. Integrated weed management. I. Key factors affecting crop weed balance. *Philippine Weed Science Bulletin* 1: 14-36.

Barbosa, P., ed. 1998. *Conservation biological control.* Academic, San Diego, CA.

Barney, R. J., W. O. Lamp, E. J. Armbrust, and G. Kapusta. 1984. Insect predator community and its response to weed management in spring-planted alfalfa. *Protection Ecology* 6: 23-33.

Baudry, J. 1984. Effects of landscape structure on biological communities: The case of heterogenous network landscapes. In *Methodology in landscape ecological research and planning,* Volume I, pp. 55-65. J. Brandt and P. Agger, eds. Roskilde University Center, Roskilde, Denmark.

Beer, J., R. Muschler, D. Kass, and E. Somarriba. 1997. Shade management in coffee and cacao plantations. American Society of Agronomy. (Symposium on Tropical Agroforestry Indianapolis, Indiana, November 5, 1996.) In *Forestry sciences: Directions in tropical agroforestry research,* pp. 139-164. P. K. R. Nair and C. R. Latt, eds. Kluwer Academic Publishers, Dordrecht, Netherlands.

Beets, W. C. 1990. *Raising and sustaining productivity of small holder farming systems in the tropics.* AgBe Publishing, Alkmaar, Holland.

Bendixen, L. E. and D. J. Horn. 1981. *An annotated bibliography of weeds as reservoirs for organisms affecting crops. III. Insects.* Agricultural Research and Development Center, Wooster, OH.

Bentley, S. and J. B. Whittaker. 1979. Effects of grazing by a chrysomelid beetle, *Gastrophysa viridula,* on competition between *Rumex obtusifolius* and *Rumex crispus. Journal of Ecology* 67: 79-90.

Bhatnagar, V S. and J. C. Davies. 1981. Pest management in intercrop subsistence farming. *Proceedings Int. Workshop on Intercropping.* ICRISAT, Patancheru, India.

Bigger, M. 1981. Observations on the insect fauna of shaded and unshaded Amelonado cocoa. *Bulletin of Entomological Research* 71 (1): 107-119.

Bobb, M. L. 1939. Parasites of the oriental fruit moth in Virginia. *Journal of Economic Entomology* 32: 605.

Boller, E. F. 1992. The role of integrated pest management in integrated production of viticulture in Europe. In *Brighton Crop Protection Conference,* pp. 499-506. British Crop Protection Council, Brighton, England.

Brown, L. R. and J. E. Young. 1990. Feeding the world in the nineties. In *State of the world,* pp. 59-78. L. R. Brown et al., eds. W. W Norton & Co., New York.

Brush, S. B. 1982. The natural and human environment of the central Andes. *Mountain Research and Development* 2: 14-38.

Buchanan, G. A. 1977. Weed biology and competition. In *Research methods in weed science*, pp. 25-41. B. Truelove, ed. Southern Weed Science Society, Auburn, AL.

Bugg, R. L. and J. D. Dutcher. 1989. Warm-season cover crops for pecan orchards: Horticultural and entomological implications. *Biological Agriculture and Horticulture* 6: 123-148.

Bugg, R. L. and C. Waddington. 1994. Using cover crops to manage arthropod pests of orchards: A review. *Agriculture, Ecosystems and Environment* 50: 11-28.

Burn, A. J. 1987. Cereal crops. In *Integrated pest management*, pp. 209-256. A. J. Burn, T. H. Coaker, and P. C. Jepson, eds. Academic Press, London.

Burn, A. J., T. H. Coaker, and P. C. Jepson, eds. 1987. *Integrated pest management.* Academic Press, London.

Buschman, L. L., H. N. Pitre, and H. F. Hodges. 1984. Soybean cultural practices: Effects on populations of geocorids, nabids, and other soybean arthropods. *Environmental Entomology* 13: 305-317.

Buttel, F. H. 1980. Agricultural structure and rural ecology: Toward a political economy of rural development. *Sociologia Ruralis* 20: 44-62.

Campbell, R. 1989. *Biological control of microbial plant pathogens.* Cambridge University Press, Cambridge, UK.

Chiverton, P. A. 1989. The creation of within-field overwintering sites for natural enemies of cereal aphids. In *Brighton Crop Protection Conference—Weeds*, pp. 1093-1096. British Crop Protection Council, Farn, Surrey.

Chiverton, P. A. and N. W. Sotherton. 1991. The effects on beneficial arthropods of the exclusion of herbicides from cereal crop edges. *Journal of Applied Ecology* 28: 1027-1039.

Churnakova, B. M. 1960. Supplementary feeding as a factor increasing the activity of parasites of harmful insects. *Trudy Vsesoyuznogo Nauchno-issledovatelscogo Instituta Zashchity Rastenii* 15: 57-70.

Coll, M. 1998. Parasitoid activity and plant species composition in intercropped systems. In *Enhancing biological control*. C. Pickett and R. Bugg, eds. University of California Press, Berkeley.

Coll, M. and D. G. Botrell. 1994. Effect of nonhost plants on an insect herbivore in diverse habitats. *Ecology* 75: 723-731.

Coll, M. and D. G. Botrell. 1996. Movement of an insect parasitoid in simple and diverse plant assemblages. *Ecological Entomology* 21: 141-149.

Collins, F. L. and S. J. Johnson. 1985. Reproductive response of caged adult velvetbean caterpillar and soybean looper to the presence of weeds. *Agriculture, Ecosystems and Environment* 14: 139-149.

Conway, G. R. 1994. Sustainability in agricultural development. *Journal for Farming Systems and Research Extension* 4: 1-14.

Conway, G. R. and J. N. Pretty. 1991. *Unwelcome harvest: Agriculture and pollution.* Earthscan Publications, London.

Coombes, D. S and N. W. Sotherton. 1986. The dispersal and distribution of polyphagous predatory Coleoptera in cereals. *Annals of Applied Biology* 108: 461-474.

Corbett, A. and R. E. Plant. 1993. Role of movement in the response of natural enemies to agroecosystem diversification: A theoretical evaluation. *Environmental Entomology* 22: 519-531.

Corbett, A. and J. A. Rosenheim. 1996. Impact of a natural enemy overwintering refuge and its interaction with the surrounding landscape. *Ecological Entomology* 21: 155-164.

Costello, M. J. and M. A. Altieri. 1994. Living mulches suppress aphids in broccoli. *California Agriculture* 48: 24-28.

Costello, M. J. and M. A. Altieri. 1995. Abundance, growth rate and parasitism of *Brevicoryne brassicae* and *Myzus persicae* (Homoptera: Aphididae) on broccoli grown in living mulches. *Agriculture, Ecosystems and Environment* 52: 187-196.

Costello, M. J. and K. M. Daane. 1998. Influence of groundcover on spider populations in a table grape vineyard. *Ecological Entomology* 23: 33-40.

Croft, B. A. 1975. Tree fruit pest management. In *Introduction to insect pest management,* pp. 471-507. R. L. Metcalf and W. H. Luckmann, eds. J. Wiley & Sons, New York.

Croft, B. A. and S. C. Hoyt. 1983. *Integrated management of insect pests of pome and stone fruits.* J. Wiley & Sons, New York.

Cromartie, W. J. 1981. The environmental control of insects using crop diversity. In *CRC handbook of pest management in agriculture,* Volume 1, pp. 223-251. D. Pimentel, ed. CRC Press, Boca Raton, FL.

Daane, K. M. and M. J. Costello. 1998. Can cover crops reduce leafhopper abundance in vineyards? *California Agriculture* 52: 27-32.

Daane, K. M., M. J. Costello, G. Y. Yakota, and W. J. Benteley. 1998. Can we manipulate leafhopper densities with management practices? *Grape Grower* 30: 18-36.

Dambach, C. A. 1948. *Ecology of crop field borders.* Ohio State University Press, Columbus, OH.

De Loach, C. J. 1970. The effect of habitat diversity on predation. *Proceedings of Tall Timbers Conference on Ecological Animal Control by Habitat Management* 2: 223-241.

Dempster, J. P. 1969. Some effects of weed control on the numbers of the small cabbage white (*Pieris rapae* L.) on brussels sprouts. *Journal of Applied Ecology* 6: 339-405.

Dempster, J. P. and T. H. Coaker. 1974. Diversification of crop ecosystems as a means of controlling pests. In *Biology in pest and disease control,* pp. 106-114. D. P. Jones and M. E. Solomon, eds. Wiley & Sons, New York.

Dennis, P. and G. L. A. Fry. 1992. Field margins: Can they enhance natural enemy population densities and general arthropod diversity on farms? *Agriculture, Ecosystems and Environment* 40: 95-115.

Doutt, R. L. and J. Nakata. 1973. The *Rubus* leafhopper and its egg parasitoid: An endemic biotic system useful in grape-pest management. *Environmental Entomology* 2: 381-386.

Duelli, P., M. Studer, I. Marchand, and S. Jakob. 1990. Population movements of arthropods between natural and cultivated areas. *Biological Conservation* 54: 193-207.

Dutcher, J. 1998. Conservation of Aphidophaga in pecan orchards. In *Conservation biological control*, pp. 37-43. P. Barbosa, ed. Academic Press, San Diego, CA.

Edland, T. 1995. Integrated pest management in fruit orchards. In *Biological control: Benefits and risks*, pp. 97-105. H. M. T. Hokkanen and J. M. Lynch, eds. Cambridge University Press, Cambridge, UK.

El Titi, A. 1986. Management of cereal pests and disease in integrated farming systems. *Proceedings of the 1986 British Crop Protection Conference—Pests and Diseases*, pp. 147-156. British Crop Protection Council, Brighton, England.

Ewel, J. J. 1986. Designing agricultural ecosystems for the humid tropics. *Annual Review Ecology and Systematics* 17: 245-271.

Ewel, J. J. 1999. Natural systems as models for the design of sustainable systems of land use. *Agroforestry Systems* 45: 1-21.

Ewel, J., F. Benedict, C. Berish, and B. Brown. 1982. Leaf area, light transmission, roots and leaf damage in nine tropical plant communities. *Agroecosystems* 7: 305-326.

Feeny, P. 1976. Plant apparency and chemical defense. *Recent Advances in Phytochemistry* 10: 1-49.

Feeny, P. 1977. Defensive ecology of the Cruciferae. *Annals of Missouri Botanical Garden* 64: 221-234.

Finch, C. V. and C. W. Sharp. 1976. *Cover crops in California orchards and vineyards*. USDA Soil Conservation Service, USDA, Washington, DC.

Finch, S. and R. H. Collier. 2000. Host-plant selection by insects: A theory based on "appropriate/inappropriate landings" by pest insects of cruciferous plants. *Entomologia Experimentalis et Applicata* 96 (2): 91-102.

Finch, S. and M. Kienegger. 1997. A behavioural study to help clarify how undersowing with clover affects host-plant selection by pests insects of brassica crops. *Entomologia Experimentalis et Applicata* 84 (2): 165-172.

Flaherty, D. 1969. Ecosystem trophic complexity and the Willamette mite, *Eotetranychus willamettei* (Acarine: Tetranychidae) densities. *Ecology* 50: 911-916.

Flint, M. L. and P. A. Roberts. 1988. Using crop diversity to manage pest problems: some California examples. *American Journal of Alternative Agriculture* 3: 164-167.

Fowler, C. and P. Mooney. 1990. *Shattering. Food, politics and the loss of genetic diversity*. The University of Arizona Press, Tucson, AZ.

Francis C. A. 1986. *Multiple cropping systems*. Macmillan, New York.

Francis, C. A. 1990. *Sustainable agriculture in temperate zones.* Wiley & Sons, New York.

Francis, C. A., C. A. Flor, and S. R. Temple. 1976. Adapting varieties for intercropped systems in the tropics. In *Multiple cropping,* pp. 235-254. ASA Special Publication No. 27. G. B. Triplett, P. A. Sanchez, and R. I. Papendick, eds. American Society of Agronomy, Madison, WI.

Frank, T. and W. Nentwig. 1995. Ground dwelling spiders (Araneae) in sown weed strips and adjacent fields. *Acta Oecologia* 16: 179-193.

Fry, G. 1995. Landscape ecology of insect movement in arable ecosystems. In *Ecology and integrated farming systems,* pp. 177-202. D. M. Glen, M. P. Greaus, and H. M. Anderson, eds. John Wiley & Sons, Briston, UK.

Fye, A. E. 1983. Cover crop manipulation for building pear psylla (Homoptera: Psyllidae) predator population in pear orchards, *Journal of Economic Entomology* 76: 306-310.

Garcia, M. A. and M. A. Altieri. 1992. Explaining differences in fleabeetle *Phyllotreta cruciferae* Goeze densities in simple and mixed broccoli cropping systems as a function of individual behavior. *Entomologia Experimentalis et Applicata* 62: 201-209.

Girma, H., M.R. Rao, and S. Sithanantham. 2000. Insect pests and beneficial arthropod populations under different hedgerow intercropping systems in semiarid Kenya. *Agroforestry Systems* 50 (3): 279-292.

Gliessman, S.R. 1999. *Agroecology: Ecological processes in agriculture.* Ann Arbor Press, Ann Arbor, MI.

Gliessman, S. R. and M. A. Amador. 1980. Ecological aspects of production in traditional agroecosystems in the humid lowland tropics of Mexico. In *Tropical ecology and development,* pp. 601-608. J. I. Furtado, ed. ISTE, Kuala Lampur.

Gold, C. S. 1987. Crop diversification and tropical herbivores: Effects of intercropping and mixed varieties on the cassava whiteflies, *Aleurotrachelus socialis* and *Rialeurodes variabilis* in Colombia. Unpublished dissertation. University of California at Berkeley.

Gold, C. S., M. A. Altieri, and A. C. Bellotti. 1989a. The effects of intercropping and mixed varieties of predators and parasitoids of cassava whiteflies (Hemiptera: Aleyrodidae) in Columbia. *Bulletin of Entomological Research* 79: 115-121.

Gold, C. S., M. A. Altieri, and A. C. Bellotti. 1989b. Relative oviposition rates of the cassava hornworm, *Erinnys ello* (Lep.: Sphingidae) and accompanying parasitism by *Telenomus sphingis* (Hym.: Scelionidae) on upper and lower surfaces of cassava. *Entomophaga* 34: 73-76.

Gold, C. S., M. A. Altieri, and A. C. Bellotti. 1990. Direct and residual effects of short duration intercrops on the cassava whiteflies *Aleurotrachelus socialis* and *Dialeurodes variabilis* (Homoptera: Aleyrodidae) in Colombia. *Agriculture, Ecosystems and Environment* 32: 57-67.

Goodman, D. 1975. The theory of diversity-stability relationships in ecology. *Quarterly Review of Biology* 50: 237-266.

Gould, F. 1986. Simulation models for predicting durability of insect-resistant germplasm: A deterministic diploid, two-locus model. *Environmental Entomology* 15: 1-10.

Grigg, D. B. 1974. *The agricultural systems of the world. An evolutionary approach.* Cambridge University Press, Cambridge, MA.

Groden, E. 1982. The interactions of toot maggots and two parasitoids, *Aleochara bilineata* (Gyll.) and *Aphaereta pillipes* (Say). MS thesis. Michigan State University, East Lansing, MI.

Guharay, F., J. Monterrey, D. Monterroso, and C. Staver. 2000. *Manejo integrado deplagos en el cultivo de café.* CATIE, Managua, Nicaragua.

Gurr, G. M., H. F. Van Emden, and S. D. Wratten. 1998. Habitat manipulation and natural enemy efficiency: Implications for the control of pests. In *Conservation biological control,* pp. 155-183. P. Barbosa, ed. Academic Press, New York.

Gut, L. J., C. E. Jochums, P. H. Westigard, and W. J. Liss. 1982. Variation in pear psylla (*Psylla pyricola* Foerster) densities in southern Oregon orchards and its implications. *Acta Horticulture* 124: 101-111.

Hagen, K. S. 1986. Ecosystem analysis: Plant cultivars (HPR) entomophagous species and food supplements. In *Interactions of plant resistance and parasitoids and predators of insects,* pp. 151-195. D. J. Boethal and R. D. Eikenbary, eds. Ellis Harwood, Chichester, England.

Harlan, J. R. 1975. Our vanishing genetic resources. *Science* 188: 618-622.

Hart, R. D. 1974. The design and evaluation of a bean, com and manioc polyculture cropping system for the humid tropics. PhD dissertation. University of Florida, Gainesville, FL.

Hart, R. D. 1980. *Agroecosistemas.* CATIE, Turrialba, Costa Rica.

Harwood, R. R. 1974. Farmer-oriented research aimed at crop intensification. In *Proceedings of cropping systems workshop,* IRRI, pp. 7-12. International Rice Research Institute, Los Banos, Philippines.

Harwood, R. R. 1979. *Small farm development—Understanding and improving farming systems in the humid tropics.* Westview Press, Boulder, CO.

Hasse, V. and J. A. Litsinger. 1981. The influence of vegetational diversity of host-finding and larval survivorship of the Asian corn borer, *Ostrinia furnacalis* Guenee. International Rice Research Institute Saturday Seminar, Department of Entomology. IRRI, Philippines.

Haynes, R. J. 1980. Influence of soil management practice on the orchard agroecosystem. *Agroecosystems* 6: 3-32.

Helenius, J. 1989. The influence of mixed intercropping of oats with field beans on the abundance and spatial distribution of cereal aphids (Homoptera: Aphididae). *Agriculture, Ecosystems and Environment* 25: 53-73.

Helenius, J. 1991. Insect numbers and pest damage in intercrops vs. monocrops: Concepts and evidence from a system of faba bean, oats and *Rhopalosiphum padi* (Homoptera, Aphididae). *Journal of Sustainable Agriculture* 1: 57-80.

Helenius, J. 1998. Enhancement of predation through within-field diversification. In *Enhancing biological control,* pp. 121-160. E. Pickett and R. L. Bugg, eds. University of California Press, Berkeley, CA.

Hendrix, P. F., D. A. Crossley, J. M. Blair, and D. C. Coleman. 1990. Soil biota as components of sustainable agroecosystems. In *Sustainable agricultural systems,* pp. 637-654. C. A. Edwards, ed. Soil and Water Conservation Society, Akeny, IO.

Hilbeck, A., M. Baumgartner, P. M. Fried, and F. Bigler. 1998. Effects of transgenic *Bacillus thuringiensis* corn-fed prey on mortality and development time of immature *Chrysoperla carnea* (Neuroptera: Chrysopidae). *Environmental Entomology* 27 (2): 480-487.

Hodek, I. 1973. *Biological of Coccinellidae.* Junk NX Publishers, Academia, The Hague, The Netherlands.

Hokkanen, H. M. T. 1991. Trap cropping in pest management. *Annual Review of Entomology* 36: 119-138.

Holland, J. M. and S. R. Thomas. 1996. *Phacelia tanacetifolia* flower strips: Their effect on beneficial invertebrates and gamebird chick food in an integrated farming system. *Acta Jutl.* 71: 171-182.

Hooks, C. R. R., H. R. Valenzuela, and J. Defrank. 1998. Incidence of pests and arthropod natural enemies on zucchini grown with living mulches. *Agriculture, Ecosystems and Environment* 69: 217-231.

Horowitz, M. T., T. Blumdel, G. Hertzlinger, and N. Hulin. 1962. Effects of repeated applications of ten soil-active herbicides on weed populations. *Weed Research* 14: 97-109.

Hoveland, C. S., G. G. Buchanan, and M. C. Harris. 1976. Response of weeds to soil phosphorus and potassium. *Weed Science* 24: 194-201.

Huffaker, C. B. and P. S. Messenger. 1976. *Theory and practice of biological control.* Academic Press, New York.

Idris, A. B. and E. Grafius. 1995. Wildflowers as nectar sources for *Diadegma insulare* (Hymenoptera: Ichneumonidae), a parasitoid of diamondback moth (Lepidoptera: Yponomeutidae). *Environmental Entomology* 24: 1726-1735.

Igzoburkie, M. U. 1971. Ecological balance in tropical agriculture. *Geographical Review* 61: 519-529.

Johnson, T. B., F. T. Turpin, and M. K. Bergman. 1984. Effect of foxtail infestation on corn rootworm larvae (Coleoptera: Chrysomelidae) under two corn-planting dates. *Environmental Entomology* 13: 1245-1248.

Kajak, A. and J. Lukasiewicz. 1994. Do semi-natural patches enrich crop fields with predatory epigean arthropods? *Agriculture, Ecosystems and Environment* 49: 149-161.

Kajimura, T.Y. 1995. Effect of organic rice farming on planthoppers. Reproduction of the white-backed planthopper, *Sogatella furcifera* (Homoptera Delphacidae). *Research on Population Ecology* 37: 219-224.

Kareiva, P. 1983. The influence of vegetation texture on herbivore populations: Resource concentration and herbivore movement. In *Variable plants and herbivores in natural and managed systems,* pp. 259-289. R. F. Denno and M. S. McClure, eds. Academic Press, New York.

Kareiva, P. 1986. Trivial movement and foraging by crop colonizers. In *Ecological theory and integrated pest management practice,* pp. 59-82. M. Kogan, ed. Wiley & Sons, New York.

Kemp, J. C. and G. W. Barrett. 1989. Spatial patterning: Impact of uncultivated corridors on arthropod populations within soybean agroecosystems. *Ecology* 70: 114-128.

Khan, Z. R., J. A. Pickett, J. van der Berg, and C. M. Woodcock. 2000. Exploiting chemical ecology and species diversity: Stemborer and *Striga* control for maize in Africa. *Pest Management Science* 56: 1-6.

Kido, H., D. L. Flaherty, C. E. Kennett, N. R. McCalley, and D. E. Bosch. 1981. Seeking the reasons for differences in orange tortrix infestations. *California Agriculture* 35: 27-28.

Klinger, K. 1987. Effects of margin-strips along a winter wheat field on predatory arthropods and the infestation by cereal aphids. *Journal of Applied Entomology* 104 (1): 47-58

Kloen, H. and M. A. Altieri. 1990. Effect of mustard *(Brassica hirta)* as a non-crop plant on competition and insect pests in broccoli *(Brassica oleraceae). Crop Protection* 9: 90-96.

Kowalski, R. and P. E. Visser. 1979. Nitrogen in a crop-pest interaction: Cereal aphids. In *Nitrogen as an ecological parameter,* pp. 67-74. J. A. Lee, ed. Blackwell Scientific Publishers, Oxford, UK.

Landis, D. A. and M. J. Haas. 1992. Influence of landscape structure on abundance and within field distribution of European corn borer (Lepidoptera: Pyralidae) larval parasitoids in Michigan. *Environmental Entomology* 21: 409-416.

Landis, D. A. and P. Marino. 1996a. Effect of landscape structure on parasitoid diversity and parasitism in agroecosystems. *Ecological Applications* 6 (1): 276-284.

Landis, D. A. and P. C. Marino. 1996b. Parasitoid communities of pest lepidoptera in agricultural landscapes: Theory, reality and implications for biological control. *Bulletin of the Ecological Society of America* 77 (3 Suppl. Part 2): 250. Annual Combined Meeting of the Ecological Society of America on Ecologists/Biologists as Problem Solvers, Providence, Rhode Island, August 10-14, 1996.

Landis, D. A., S. D. Wratten, and G. A. Gurr. 2000. Habitat management to conserve natural enemies of arthropod pests in agriculture. *Annual Review of Entemology* 45: 175-201.

Lasack, P. M. and L. P. Pedigo. 1986. Movement of stalk borer larvae (Lepidoptera: Noctuidae) from noncrop areas into corn. *Journal of Economic Entomology* 79: 1697-1702.

Lashomb, J. H. and Y. S. Ng. 1984. Colonization by Colorado potato beetles, *Leptinotarsa decemlineata* (Say) (Coleoptera: Chrysomelidae), in rotated and non-rotated potato fields. *Environmental Entomology* 13: 1352-1356.

Latheef, M. A. and R. D. Irwin. 1980. Effects of companionate planting on snap bean insects, *Epilachna varivestis* and *Heliothis zea*. *Environmental Entomology* 9: 195-198.

Leius, K. 1967. Influence of wild flowers on parasitism of tent caterpillar and codling moth. *Canadian Entomologist* 99: 444-446.

LeSar, C. D. and J. D. Unzicker. 1978. Soybean spiders: Species composition, population densities and vertical distribution. *Illinois Natural History Survey. Biological Notes No. 107*. Ubana, IL.

Leston, D. 1973. The ant mosaic-tropical tree crops and the limiting of plots and diseases. *PANS* 19: 311.

Letourneau, D. K. 1983. The effects of vegetational diversity on herbivorous insects and associated natural enemies: Examples from tropical and temperate agroecosystems. PhD dissertation. University of California, Berkeley, CA.

Letourneau, D. K. 1987. The enemies hypothesis: Tritrophic interaction and vegetational diversity in tropical agroecosystems. *Ecology* 68: 1616-1622.

Letourneau, D. K. and M. A. Altieri. 1983. Abundance patterns of a predator *Orius tristicolor* (Hemiptera: Anthoconidae) and its prey, *Frankliniella occidentalis* (Thysanoptera: Thripidae): Habitat attraction in polycultures versus monocultures. *Environmental Entomology* 122: 1464-1469.

Levine, E. 1985. Oviposition by the stalk borer, *Papailpema nebris* (Lepidoptera: Noctuidae) on weeds, plant debris, and cover crops in cage tests. *Journal of Economic Entomology* 78: 65-68.

Levins, R. and M. Wilson. 1979. Ecological theory and pest management. *Annual Review of Entomology* 25: 7-29.

Lewis, T. 1965. The effects of shelter on the distribution of insect pests. *Scientific Horticulture* 17: 74-84.

Liang, W. and M. Huang. 1994. Influence of citrus orchard groundcover plants on arthropod communities in China: A review. *Agriculture, Ecosystems and Environment* 50: 29-37.

Liebman, J. 1997. *Rising toxic tide: Pesticide use in California, 1991-1995*. Pesticide Action Network, San Francisco, CA.

Liebman, M. and E. R. Gallandt. 1997. Many little hammers: Ecological management of crop-weed interactions. In *Ecology in agriculture*, pp. 291-343. L. E. Jackson, ed. Academic Press, San Diego, CA.

Lincoln, C. and D. Isley. 1947. Corn as a trap crop for the cotton bollworm. *Journal of Economic Entomology* 40: 437-438.

Litsinger, J. A., V. Hasse, A. T. Barrion, and H. Schmutterer. 1991. Response of *Ostrinia furnacalis* (Guenee) (Lepidoptera: Pyralidae) to intercropping. *Environmental Entomology* 20: 988-1004.

Litsinger, J. A. and K. Moody. 1976. Integrated pest management in multiple cropping systems. In *Multiple Cropping*, pp. 293-316. R. I. Papendick, P. A. Sanchez, and G. B. Triplett, eds. Special Publication 27. American Society of Agronomy, Madison, WI.

Lys, J. A. 1994. The positive influence of strip-management on ground beetles in a cereal field: Increase, migration and over-wintering. In *Carabid beetles: Ecology and evolution*, pp. 451-455. K. Desender, M. Dufrene, M. Loreau, M. L. Luff, and J. P. Maelfait, eds. Kluwer, Dortdrecht/Boston/London.

Lys, J. A. and W. Nentwig. 1992. Augmentation of beneficial arthropods by strip-management. 4. Surface activity, movements and density of abundant carabid beetles in a cereal field. *Oecologia* 92: 373-382.

Maier, C. T. 1981. Parasitoids emerging from puparia of *Rhagoletis pomenella* (Diptera: Tephritidae) infesting hawthorn and apple in Connecticut. *Canadian Entomologist* 113: 867.

Marcovitch, S. 1935. Experimental evidence on the value of strip cropping as a method for the natural control of injurious insects, with special reference to plant lice. *Journal of Economic Entomology* 28: 26-70.

Marino, P. C. and D. L. Landis. 1996. Effect of landscape structure on parasitoid diversity and parasitism in agroecosystems. *Ecological Applications* 6: 276-284.

Marvier, M. 2001. Ecology of transgenic crops. *American Scientist* 89: 160-167.

Matteson, P. C., M. A. Altieri, and W. C. Gagne. 1984. Modification of small farmer practices for better management. *Annual Review of Entomology* 29: 383-402.

Mayse, M. A. 1983. Cultural control in crop fields: A habitat management technique. *Environmental Entomology* 7: 15-22.

Mayse, M. A. and P. W. Price. 1978. Seasonal development of soybean arthropod communities in east central Illinois. *Agroecosystems* 4: 387-405.

McClure, M. 1982. Factors affecting colonization of an orchard by leafhopper (Homoptera: Cicadellidae) vectors of peach x-disease. *Environmental Entomology* 11: 695.

McNeely, J. A., K. R. Miller, W. V. Reid, R. A. Mittermeier, and T. B. Werner. 1990. *Conserving the world's biological diversity.* International Union for Conservation of Nature and Natural Resources. World Resource Institute, Conservation International, World Wildlife Fund, World Bank, Washington, DC.

Michigan State University Extension. 2000. *Michigan field crop pest ecology and management.* Bulletin E2704. Author, Lansing, MI.

Monteith, L. G. 1960. Influence of plants other than the food plants of their host on host-finding by tachinid parasites. *Canadian Entomologist* 92: 641-652.

Mortinson, T. E., J. P. Nyrop, and C. J. Eckenroad. 1988. Dispersal of the onion fly (Diptera Anthomyildae) and larval damage in rotated onion field. *Journal of Economic Entomology* 81: 509-514.

Murdoch, W. W. 1975. Diversity, stability, complexity and pest control. *Journal of Applied Ecology* 12: 745-807.

Murphy, B.C., J. A. Rosenheim, R.V. Dowell, and J. Granett. 1998. Habitat diversification tactic for improving biological control: Parasitism of the western grape leafhopper. *Entomologia Experimentalis et Applicata* 87 (3): 225-235.

Murphy, B. C., J. A. Rosenheim, and J. Granett. 1996. Habitat diversification for improving biological control: Abundance of *Anagrus epos* (Hymenoptera: Mymaridae) in grape vineyards. *Environmental Entomology* 25 (2): 495-504.

Myers, N. 1984. *The primary source: Tropical forests and our future.* W. W. Norton, New York.

Nafus, D. and I. Schreiner. 1986. Intercropping maize and sweet potatoes. Effects on parasitization of *Ostriniafurnacalis* eggs by *Trichogramma chilonis. Agriculture, Ecosystem and Environment* 15: 189-200.

Nair, P. K. 1993. *An introduction to agroforestry.* Kluwer Academic Publishers. Dordrecht, Netherlands.

Nascimento, A. S., A. L. M. Mesquita, and R. C. Caldas. 1986. Population fluctuation of the citrus borer, *Cratosomus flavofasciatus* Guerin, 1844 (Coleoptera: Curculionidae), on the trap plant, *Cordia verbenacea* (Boraginaceae). *Anais da Sociedade Entomologica do Brasil* 15: 125-134.

National Academy of Sciences. 1969. *Principles of plant and animal control.* Volume 3. Insect pest management and control. National Academy of Science, Washington, DC.

National Academy of Sciences. 1972. *Genetic vulnerability of major crops.* National Academy of Science, Washington, DC.

Nentwig, W. 1998. Augmentation of beneficial arthropods by strip management. 1. Succession of predaceous arthropods and long-term change in the ratio of phytophagous and predaceous species in a meadow. *Oecologia* 76: 597-606.

Nettles, W. C. 1979. *Eucelatoria* sp. females: Factors influencing response to cotton and okra plants. *Environmental Entomology* 8: 619-623.

Nicholls, C. I., M. P. Parrella, and M. A. Altieri. 2000. Reducing the abundance of leafhoppers and thrips in a northern California organic vineyard through maintenance of full season floral diversity with summer cover crops. *Agricultural and Forest Entomology* 2: 107-113.

Nicholls, C. I., M. P. Parrella, and M. A. Altieri. 2001. The effects of a vegetational corridor on the abundance and dispersal of insect biodiversity within a northern California organic vineyard. *Landscape Ecology* 16: 133-146.

Nordlund, D. A., R. B. Chalfant, and W. J. Lewis. 1984. Arthropod populations, yield and damage in monocultures and polycultures of corn, beans and tomatoes. *Agriculture, Ecosystems and Environment* 11: 353-367.

Norman, M. J. T. 1979. *Annual cropping systems in the tropics.* University Presses of Florida, Gainesville, FL.

Norris, R. R. 1982. Interactions between weeds and other pests in the agroecosystem. In *Biometeorology in integrated pest management,* pp. 343-406. J. L. Hatfield and I. J. Thomason, eds. Academic Press, New York.

O'Connor, B. A. 1950. Premature nutfall of coconuts in the British Solomon Islands Protectorate. *Agriculture Journal,* Fiji Department of Agriculture 21: 1-22.

Palada, M. C., S. Ganser, R. Hofstetter, B. Volak, and M. Culik. 1983. Associations of interseeded legume cover crops and annual row crops in year-round cropping systems. In *Experimentally sound agriculture*, pp. 193-214. N. Lockeretz, ed. Praeger, New York.

Palti, J. 1981. *Cultural practices and infectious crop diseases.* Springer, New York.

Paoletti, M. G., B. R. Stinner, and G. G. Lorenzoni. 1989. *Agricultural ecology and environment.* Elsevier, Amsterdam.

Peng, R. K., L. D. Incoll, S. L. Sutton, C. Wright, and A. Chadwick. 1993. Diversity of airborne arthropods in a silvoarable agroforestry system. *Journal of Applied Ecology* 30 (3): 551-562.

Peppers, B. B. and B. F. Driggers. 1934. Non-economic insects as intermediate hosts of parasites of the oriental fruit moth. *Annals of the Entomological Society of America* 27: 593-598.

Perfect, T. J. 1991. Biodiversity and tropical pest management. In *The Biodiversity of microorganisms and invertebrates: Its role in sustainable agriculture*, pp. 145-148. D. L. Hawksworth, ed. CAB International, Wallingford, UK.

Perfecto, I. 1995. Biodiversity and the transformation of a tropical agroecosystem: Ants in coffee plantations. *Ecological Applications* 5 (4): 1084-1097.

Perfecto, I. and J. H. Vandermeer. 1996. Microclimatic changes and the indirect loss of ant diversity in a tropical agroecosystem. *Oecologia* 108: 577-582.

Perrin, R. M. 1975. The role of the perennial stinging nettle *Urtica dioica* as a reservoir of beneficial natural enemies. *Annals of Applied Biology* 81: 289-297.

Perrin, R. M. 1977. Pest management in multiple cropping systems. *Agroecosystems* 3: 93-118.

Perrin, R. M. 1980. The role of environmental diversity in crop protection. *Protection Ecology* 2: 77-114.

Perrin, R. M. and M. L. Phillips. 1978. Some effects of mixed cropping on the population dynamics of insect pests. *Entomologia Experimentalis et Applicata* 24: 385-393.

Peterson, P. 1926. Oriental fruit moth damage in cultivated and uncultivated orchards. *Proceedings of the Annual Meeting of the New Jersey State Horticultural Society* 3: 83-86.

Phelan, P. L., J. F. Mason, and B. R. Stinner. 1995. Soil-fertility management and host preference by European corn borer, *Ostrinia nubilalis* (Huebner), on *Zea mays* L.: A comparison of organic and conventional chemical farming. *Agriculture, Ecosystems and Environment* 56 (1): 1-8.

Pickett, C.H. and R.L. Bugg. 1998. *Enhancing biological control: Habitat management to promote natural enemies of agricultural pests.* University of California Press, Berkeley, CA.

Pierce, W. D., R. A. Cushman, and C. E. Hood. 1912. *The insect enemies of the cotton boll weevil.* Bureau of Entomology Bulletin No. 100, USDA, Washington, DC.

Pimentel, D. 1961. Species diversity and insect population outbreaks. *Annals of Entomological Society of America* 54: 76-86.

Pimentel, D., D. Andow, R. Dyson-Hudson, D. Gallahan, S. Jacobson, M. Irish, S. Kroop, A. Moss, I. Schreiner, M. Shepard, et al. 1980. Environmental and social costs of pesticides: A preliminary assessment. *Oikos* 34: 126-140.

Pollard, E. 1968. Hedges IV. A comparison between the carabidae of a hedge and field site and those of a woodland glade. *Journal of Applied Ecology* 5: 649-657.

Powell, W. 1986. Enhancing parasitoid activity in crops. In *Insect parasitoids*, pp. 319-335. J. Waage and D. Greathead, eds. Academic Press, London.

Power, A. G. 1987. Plant community diversity, herbivore movement, and an insect-transmitted disease of maize. *Ecology* 68: 1658-1669.

Pretty, J. N. 1995. *Regenerating agriculture: Policies and practice for sustainability and self-reliance.* Earthscan Publications, London.

Pretty, J. N. 1997. The sustainable intensification of agriculture. *National Resources Forum* 21: 247-256.

Pretty, J. N. and R. Hine. 2000. *Feeding the world with sustainable agriculture: A summary of new evidence.* Final report from "SAFE-World" Research Project. University of Essex, England.

Price, P. 1976. Colonization of crops by arthropods: Non-equilibrium communities in soybean fields. *Environmental Entomology* 5: 605-611.

Price, P. W., C. E. Bouton, P. Gross, B. A. McPherson, J. N. Thompson, and A. E. Weise. 1980. Interactions among three trophic levels: Influence of plants on interaction between insect herbivores and natural enemies. *Annual Review of Ecology* 11: 41-60.

Prokopy, R.C. 1994. Integration in orchard pest and habitat management: A review. *Agriculture, Ecosystems and Environment* 50: 1-10.

Puvuk, D. M. and B. R. Stinner. 1992. Influence of weed communities in corn plantings on parasitism of *Ostrinia nubilalis* (Lepidoptera: Pyralidae) by *Erioborus terebrans* (Hymenoptera: Ichneumonidae). *Biological Control* 2: 312-316.

Rabb, R. L. 1978. A sharp focus on insect populations and pest management from a wide area view. *Bulletin of the Entomological Society of America* 24: 55-60.

Rabb, R. L., R. E. Stinner, and R. van den Bosch. 1976. Conservation and augmentation of natural enemies. In *Theory and practice of biological control*, pp. 233-254. C. B. Huffaker and P. Messenger, eds. Academic Press, New York.

Rao, M. R., M.P. Singh, and R. Day. 2000. Insect pest problems in tropical agroforestry systems: Contributory factors and strategies for management. *Agroforestry Systems* 50 (3): 243-277.

Read, D. P., P. P. Feeny, and R. B. Root. 1970. Habitat selection by the aphid parasite *Diaeretiella rapae* (Hymenoptera: Braconidae) and hyperparasite *Charips brassica* (Hymenoptera: Cynipidae). *Canadian Entomologist* 102: 1567-1578.

Reganold, J. P., J. D. Glover, P. K. Andrews, and H. R. Hinman. 2001. Sustainability of three apple production systems. *Nature* 410: 926-930.

Rhoades, D. F. and R. G. Cates. 1976. Toward a general theory of plant anti-herbivore chemistry. *Recent Advances in Phytochemistry* 10: 168-213.

Risch, S. J. 1980. The population dynamics of several herbivorous beetles in a tropical agroecosystem: the effect of intercropping corn, beans and squash in Costa Rica. *Journal of Applied Ecology* 17: 593-612.

Risch, S. J. 1981. Insect herbivore abundance in tropical monocultures and polycultures: An experimental test of two hypotheses. *Ecology* 62: 1325-1340.

Risch, S. J. 1983. Intercropping as a cultural pest control: Prospects and limitations. *Environmental Management* 7: 9-14.

Risch, S. J., D. Andow, and M. A. Altieri. 1983. Agroecosystem diversity and pest control: Data, tentative conclusions, and new research directions. *Environmental Entomology* 12 (3): 625-629.

Rissler, J. and M. Mellon. 1996. *The ecological risks of engineered crops.* MIT Press, Cambridge, MA.

Robinson, R. A. 1996. *Return to resistance: Breeding crops to reduce pesticide resistance.* Ag. Access. Davis, CA.

Robinson, R. R., J. H. Young, and R. D. Morrison. 1972. Strip-cropping effects on abundance of predatory and harmful cotton insects in Oklahoma. *Environmental Entomology* 1: 145-149.

Root, R. B. 1973. Organization of a plant-arthropod association in simple and diverse habitats: The fauna of collards *(Brassica oleracea). Ecological Monographs* 43: 95-124.

Root, R. B. 1975. Some consequences of ecosystem texture. In *Ecosystem analysis and prediction.* S. A. Levin, ed. Penn State University Press, Philadelphia, PA.

Rosset, P., J. Vandermeer, M. Cano, P. G. Varela, A. Snook, and C. Hellpap. 1985. El Frijol como cultivo trampa para el combate de *Spodoptera sunia* Guenee (Lepidoptera: Noctuidae) en plantulas de tomate. *Agronomia Costarricense* 9: 99-102.

Russell, E. P. 1989. Enemies hypothesis: A review of the effect of vegetational diversity on predatory insects and parasitoids. *Environmental Entomology* 18: 590-599.

Sanchez, P. A. 1995. Science in agroforestry. *Agroforestry Systems* 30: 5-55.

Schellhorn, N. A. and V. L. Sork. 1997. The impact of weed diversity on insect population dynamics and crop yield in collards, *Brassica oleracea* (Brassicaceae). *Oecologia* 111 (2): 233-240.

Schoonhoven, A. von, C. Cardona, J. Garcia, and R. Garzon. 1981. Effect of weed covers on *Empoasca kraemeri* Ross and Moore. Populations and dry bean yields. *Environmental Entomology* 10: 901-907.

Schroth, G., U. Krauss, L. Gasparotto, J. Aguilar, A. Duarte, and K. Vohland. 2000. Pests and diseases in agroforestry systems of the humid tropics. *Agroforestry Systems* 50 (3): 199-241.

Sengonca, C. and B. Frings. 1988. The influence of *Phacelia tanacetifolia* to pests and beneficial insects in sugar beet plots. *Pedobiologia* 32 (5-6): 311-316.

Settle, W. H., L. T. Wilson, D. L. Flaherty, and G. M. English-Loeb. 1986. The variegated leafhopper, as increasing pest of grapes. *California Agriculture* 40: 30-32.

Shahjahan, M. and A. S. Streams. 1973. Plant effects on host-finding by *Leiophron pseudopallipes* (Hymenoptera: Braconidae), a parasitoid of the tarnished plant bug. *Environmental Entomology* 2: 921-925.

Sheehan, W. 1986. Response by specialist and generalist natural enemies to agroecosystem diversification: A selective review. *Environmental Entomology* 15: 456-461.

Slosser, J. E. and E. P. Boring III. 1980. Shelterbelts and bollweevils: A control strategy based in management of overwintering habitat. *Environmental Entomology* 9: 1-6.

Slosser, J. E., R. F. Gewin, I. R. Price, L. J. Meinke, and J. R. Bryson. 1984. Potential of shelterbelt management for boll weevil (Col: Curculionidae) control in the Texas Rolling Plains. *Journal of Economic Entomology* 77: 377-385.

Sluss, R. R. 1967. Population dynamics of the walnut aphid *Chromaphis juglandicola* (Kalt.) in northern California. *Ecology* 48: 41-58.

Smith, H. A. and R. McSorley. 2000. Intercropping and pest management: A review of major concepts. *American Entomologist* 46: 154-161.

Smith, J. G. 1969. Some effects of crop background on populations of aphids and their natural enemies on brussels sprouts. *Annals of Applied Biology* 63: 326-330.

Smith, J.G. 1976a. Influence of crop background on aphids and other phytophagous insects on brussels sprouts. *Annals of Applied Biology* 83: 1-13.

Smith, J.G. 1976b. Influence of crop background on natural enemies of aphids on brussels sprouts. *Annals of Applied Biology* 83: 19-22.

Smith, M. W., D. C. Arnold, R. D. Eikenbary, N. R. Rice, A. Shiferaw, B. S. Cheary, and B. L. Carroll. 1996. Influence of groundcover on beneficial arthropods in pecan. *Biological Control* 6: 164-176.

Smith, R. F. and H. T. Reynolds. 1972. Effects of manipulation on cotton agroecosystems on insect pest populations. In *The careless technology,* pp. 183-192. T. Farvar and J. P. Martin, eds. Natural History Press, New York.

Solomon, M. G. 1981. Windbreaks as a source of orchard pests and predators. In *Pests, pathogens and vegetation: The role of weeds and wild plants in the ecology of crop pests and diseases,* pp. 273-283. J. M. Thresh, ed. Pitman, Boston, MA.

Sotherton, N. W. 1984. The distribution and abundance of predatory arthropod overwintering on farmland. *Annals of Applied Biology* 105: 423-424.

Soule, J. D. and J. K. Piper. 1992. *Farming in nature's image.* Island Press, Washington, D.C.

Southwood, T. R. E. and M. J. Way. 1970. Ecological background to pest management. In *Concepts of pest management,* pp. 7-13. R. L. Rabb and F. E. Guthrie, eds. North Carolina State University, Raleigh, NC.

Speight, M. R. 1983. The potential of ecosystem management for pest control. *Agriculture, Ecosystems and Environment* 10: 183-199.

Speight, M. and J. H. Lawton. 1976. The influence of weed cover on the mortality imposed on artificial prey by predatory ground beetles in cereal fields. *Oecologia* 23: 211-223.

Srinivasan, K. and P. N. K. Moorthy. 1991. Indian mustard as a trap crop for management of major lepidopterous pests on cabbage. *Tropical Pest Management* 37: 26-32.

Stamps, W. T. and M. J. Linit. 1997. Plant diversity and arthropod communities: Implications for temperate agroforestry. *Agroforestry Systems* 39 (1): 73-89.

Staver, C., F. Guharay, D. Monteroso, and R.G. Muschler. 2001. Designing pest-suppresive multistrata perennial crop systems: Shade-grown coffee in Central America. *Agroforestry Systems* 53: 151-170.

Stephens, C. S. 1984. Ecological upset and recuperation of natural control of insect pests in some Costa Rican banana plantations. *Turrialba* 34: 101-105.

Stern, V. 1981. Environmental control of insects using trap crops, sanitation, prevention and harvesting. In *CRC Handbook of Pest Management in Agriculture*. Volume 1, pp. 199-207. D. Pimentel, ed. CRC Press, Boca Raton, FL.

Stern, V. M. 1979. Interplanting alfalfa in cotton to control lygus bugs and other insect pests. *Proceedings of Tall Timbers Conference on Ecological Animal Control by Habitat Management* 1: 21-26.

Stinner, R. E., C. S. Barfield, J. L. Stimac, and L. Dohse. 1983. Dispersal movement of insect pests. *Annual Review of Entomology* 28: 319-335.

Strong, D. R., J. H. Lawton, and R. Southwood. 1984. *Insects on plants: Community patterns and mechanisms*. Harvard University Press, London.

Swift, M. J. and J. M. Anderson. 1993. Biodiversity and ecosystem function in agricultural systems. In *Biodiversity and ecosystem function*, pp. 15-22. E. D. Scholze and H. Mooney, eds. Spriger, Berlin, Germany.

Syme, P. D. 1975. The effects of flowers on the longevity and fecundity of two native parasites of the European pine shoot moth in Ontario. *Environmental Entomology* 4: 337-346.

Tahvanainen, J. O. and R. B. Root. 1972. The influence of vegetational diversity on the population ecology of a specialized herbivore, *Phyllotreta cruciferae* (Coleoptera: Chrysomehdae). *Oecologia* 10: 321-346.

Telenga, N. A. 1958. Biological method of pest control in crops and forest plants in the USSR. In *Report of the Soviet Delegation*, pp. 1-15. Ninth International Conference on Quarantine and Plant Protection, Moscow.

Theunissen, J., C. J. H. Booij, and L. A. Lotz. 1995. Effects of intercropping white cabbage with clovers on pest infestation and yield. *Entomologia Experimentalis et Applicata* 74: 7-16.

Theunissen, J. and H. den Ouden. 1980. Effects of intercropping with *Spergula arvensis* on pests of brussels sprouts. *Entomologia Experimentalis et Applicata* 22: 260-268.

Thiele, H. 1977. *Carabid beetles in their environments*. SpringerVerlag, New York.

Thies, C. and T. Tscharntke. 1999. Landscape structure and biological control in agroecosystems. *Science* 285: 893-895.

Thomas, M. B. and S. D. Wratten. 1990. Ecosystem diversification to encourage natural enemies of cereal aphids. In *Pests and diseases*, pp. 691-696. British Crop Protection Conference, Brighton, England.

Thomas, M. B., S. D. Wratten, and N. W. Sotherton. 1992. Creation of "island" habitats in farmland to manipulate populations of beneficial arthropods: Predator densities and species composition. *Journal of Applied Ecology* 29: 524-531.

Thresh, J. M. 1981. *Pests, pathogens and vegetation: The role of weeds and wild plants in the ecology of crop pests and diseases.* Pitman, Boston, MA.

Thrupp, L. A. 1997. *Linking biodiversity and agriculture: Challenges and opportunities for sustainable food security.* World Resources Institute, Washington, DC.

Tilman, D., D. Wedin, and J. Knops. 1996. Productivity and sustainability influenced by biodiversity in grassland ecosystems. *Nature* 379: 718-720.

Toledo, V. M., J. Cararbias, C. Mapes, and C. Toledo. 1985. *Ecologia y Autosuficiencia Alimentaría.* Siglo Veintiuno, Mexico City, Mexico.

Tonhasca, A. 1993. Effects of agroecosystem diversification on natural enemies of soybean herbivores. *Entomologia Experimentalis et Applicata* 69: 83-90.

Topham, M. and J. W. Beardsley. 1975. An influence of nectar source plants on the New Guinea sugarcane weevil parasite, *Lixophaga sphenophori* (Villeneuve). *Proceedings of the Hawaiian Entomological Society* 22: 145-155.

Trenbath, B. R. 1976. Plant interactions in mixed crop communities. In *Multiple cropping,* pp. 129-169. ASA Special Publication No. 27. G. B. Triplett, P. A. Sanchez, and R. J. Papendick, eds. American Society of Agronomy, Madison, WI.

Trujillo-Arriaga, J. and M. A. Altieri. 1990. A comparison of aphidophagous arthropods on maize polycultures and monocultures in Central Mexico. *Agriculture, Ecosystems and Environment* 31: 337-349.

Turnbull, A. L. 1969. The ecological role of pest populations. *Proceedings of Tall Timbers Conference on Ecological Animal Control by Habitat Management* 1: 219-232.

U.S. Department of Agriculture (USDA). 1973. Monocultures in agriculture, causes and problems. *Report of the Task Force on Spatial Heterogeneity in agricultural landscapes and enterprises.* U.S. Government Printing Office, Washington, DC.

Uvah, I. I. I. and T. H. Coaker. 1984. Effect of mixed cropping on some insect pests of carrots and onions. *Entomologia Experimentalie et Applicata* 36: 159-167.

van den Bosch, R. and V. M. Stern. 1969. The effects of harvesting practices on insect populations in alfalfa. In *Proceedings of the tall timbers conference on ecological animal control by habitat management, Volume 1,* pp. 47-54. Tall Timbers Research Station, Tallahassee, FL.

van den Bosch, R. and A. D. Telford. 1964. Environmental modification and biological control. In *Biological control of insect pests and weeds,* pp. 459-488. P. DeBach, ed. Chapman and Hall, London.

Van Driesche, R. G. and T. S. Bellows Jr. 1996. *Biological control.* Chapman and Hall, New York.

Van Emden, H. F. 1965a. The effect of uncultivated land on the distribution of cabbage aphid on an adjacent crop. *Journal of Applied Ecology* 2: 171-196.

Van Emden, H. F. 1965b. The role of uncultivated land in the biology of crop pests and beneficial insects. *Scientific Horticulture* 17: 121-136.

Van Emden, H. F. 1990. Plant diversity and natural enemy efficiency in agro-ecosystems. In *Critical issues in biological control*, pp. 63-80. M. Mackauer, L.E. Ehler, and J. Roland, eds. Intercept Ltd, Andover, UK.

Van Emden, H. F. and Z. T. Dabrowski. 1997. Issues of biodiversity in pest management. *Insect Science and Applications* 15: 605-620.

Van Emden, H. F. and G. F. Williams. 1974. Insect stability and diversity in agroecosystem. *Annual Review of Entomology* 19: 455-475.

van Huis, A. 1981. *Integrated pest management in the small farmer's maize crop in Nicaragua*. H. Veenman and B. V. Zonen, Wageningen, The Netherlands.

Vandermeer, J. 1981. The interference production principle: An ecological theory for agriculture. *BioScience* 31: 361-364.

Vandermeer, J. 1989. *The ecology of intercropping*. Cambridge University Press, Cambridge, UK.

Vandermeer, J. and I. Perfecto. 1995. *Breakfast of biodiversity*. Food First Books, Oakland, California.

Varchola, J. M. and J. P. Dunn. 1999. Changes in ground beetle (Coleoptera: Carabidae) assemblages in farming systems bordered by complex or simple roadside vegetation. *Agriculture, Ecosystems and Environment* 73 (1): 41-49.

Vrabel, T. E., P. L. Minnotti, and R. D. Sweet. 1980. Seeded legumes as living mulches in corn. Paper No. 764. Department of Vegetable Crops. Cornell University, Ithaca, NY.

Wainhouse, D. and T. H. Coaker. 1981. The distribution of carrot fly *(Psila rosae)* in relation to the fauna of field boundaries. In *Pests, pathogens and vegetation: The role of weeds and wild plants in the ecology of crop pests and diseases*, pp. 263-272. J. H. Thresh, ed. Pitman, Boston, MA.

Walker, R. H., M. G. Patterson, E. Hauser, D. Isenhour, J. Todd, and G. A. Buchanan. 1984. Effects of insecticide, weed-free period, and row spacing on soybean *(Glycine max)* and sicklepod *(Cassia obtusifolia)* growth. *Weed Science* 32: 702-706.

Wallin, H. 1985. Spatial and temporal distribution of some abundant carabid beetles (Coleoptera: Carabidae) in cereal fields and adjacent habitats. *Pedobiologia* 28 (1): 19-34.

Watt, R. E. F. 1973. *Principles of environmental science*. McGraw Hill, New York.

Way, M. J. 1977. Pest and disease status in mixed stands vs. monocultures: The relevance of ecosystem stability. In *Origins of pest, parasite, disease and weed problems*, pp. 127-138. J. M. Cherrett and G. R. Sagar, eds. Blackwell, Oxford, UK.

Wetzler, R. E. and S. J. Risch. 1984. Experimental studies of beetle diffusion in simple and complex crop habitats. *Journal of Animal Ecology* 53: 1-19.

William, R. D. 1981. Complementary interactions between weeds, weed control practices, and pests in horticultural cropping systems. *HortScience* 16: 508-513.

Wood, B. J. 1971. Development of integrated control programs for pests of tropical perennial crops in Malaysia. In *Biological control*, pp. 422-457. C. B. Huffaker, ed. Plenum Press, New York.

Wratten, S. D. 1987. The effectiveness of native natural enemies. In *Integrated pest management*, pp. 89-112. A. J. Burn, T. H. Coaker, and P. C. Jepson, eds. Academic Press, London.

Wratten, S.D. and H.F.Van Emden. 1995. Habitat management for enhanced activity of natural enemies of insect pests. In *Ecology and integrated farming systems*, pp. 117-145. D. M. Glen, M. P. Greaves, and H. M. Anderson, eds. John Wiley and Sons, Chichester, UK.

Wrubel, R. P. 1984. The effect of intercropping on the population dynamics of the arthropod community associated with soybean *(Glycine max)*. MS thesis. University of Virginia, Blacksburg, VA.

Wyss, E., U. Niggli, and W. Nentwig. 1995. The impact of spiders on aphid populations in a strip-managed apple orchard. *Journal of Applied Entomology* 114: 473-478.

Yan, Y. H., Y. Yi, X. G. Du, and B. G. Zhao. 1997. Conservation and augmentation of natural enemies in pest management of Chinese apple orchards. *Agriculture, Ecosystems and Environment* 62 (2-3): 253-260.

Yurjevic, A. M. 1991. The assessment of agroecological bottom up development. PhD dissertation. Latin American Studies, University of California, Berkeley, CA.

Zandstra, B. H. and P. S. Motooka. 1978. Beneficial effects of weeds in pest management—A review. *PANS* 24: 333-338.

Zimdahl, R. L. 1980. *Weed-crop competition—A review*. International Plant Protection Center, Corvallis, OR.

Index

Page numbers followed by the letter "f" indicate figures; those followed by the letter "t" indicate tables.